THE SOCIAL PSYCHOLOGY OF MILITARY SERVICE

Edited by NANCY L. GOLDMAN and DAVID R. SEGAL

SAGE RESEARCH PROGRESS SERIES ON
WAR, REVOLUTION, AND PEACEKEEPING

Volume VI (1976)

SAGE PUBLICATIONS *Beverly Hills / London*

For information address:

SAGE PUBLICATIONS, INC.
275 South Beverly Drive
Beverly Hills, California 90212

SAGE PUBLICATIONS LTD
St George's House / 44 Hatton Garden
London EC1N 8ER

Printed in the United States of America

International Standard Book Number 0-8039-0598-X (c)
0-8039-0599-8 (p)

Library of Congress Catalog Card No. 76-2111

FIRST PRINTING

CONTENTS

INTRODUCTION

The advent of an all-volunteer military force with worldwide responsibilities is a unique development in the United States and has raised a range of crucial research and policy issues. Of special importance are problems of manpower and personnel for this new type of military organization, and an understanding of new dimensions of civil-military relations as they apply in the all-volunteer period.

In order to provide information fundamental to the understanding of the all-volunteer military and to provide the necessary basic data for the development of manpower and leadership policies, a two-day research conference on the social psychology of military service was held on April 23-25, 1975, at the Center for Continuing Education, University of Chicago, with Dr. David R. Segal, then of the U.S. Army Research Institute, and Ms. Nancy L. Goldman of the University of Chicago serving as conference cochairmen.

At this forum, social scientists from a variety of academic disciplines engaged in research on the transition to an all-volunteer force, identified the social and psychological dimensions of the all-volunteer military, assessed the social character of the military, and compared the specific problems of motivation and utilization in the all-volunteer force with such problems in the selective service army.

Although much research has been conducted on military training and education, little recent work has been concerned with processes of socialization and adjustment in the training environment. In the first paper in this volume, John H. Faris addresses these issues in the context of basic combat training and cites the importance of noncommissioned officers in making men into soldiers.

Next, a set of two papers deals with adjustment to military environments in an operational rather than a training setting. Jonathan F. Borus addresses the readjustment of soldiers returning from the Vietnam theater to stateside garrison environments. Both he and the next author, Robert Jay Lifton, are concerned with identifying the institutions responsible for easing the transition of the combat soldier returning to civil society.

Adjustment to the military is potentially problematic even for personnel who are not serving in or returning from a combat theater. Because of its unique function in society—the management of organized violence—the military must be organizationally different from civilian employers. Beginning a set of papers, E.K. Eric Gunderson discusses the unique problems of sailors at sea. Paul D. Nelson reviews the literature on social-background characteristics associated with adjustment, while David G. Bowers presents comparisons of the work-related attitudes of civilians and of Navy personnel.

The role of the family is emerging as a central issue in military sociology and psychology, because the all-volunteer force is a predominantly married military population, in contrast to the predominantly bachelor force of the conscription era. M. Duncan Stanton's paper identifies some of the problems experienced by families in a military environment. To highlight both the importance and the recent increase in research in this area, we have included in this part two papers that were not presented at the Chicago conference. The first, by Nancy L. Goldman, presents important demographic data on trends in the structure of military families, highlighting the end of the bachelor army. The other paper, by Ellwyn R. Stoddard and Claude E. Cabanillas, discusses the role of the officer's wife (which is rapidly changing as a function of sex-role definitions) and puts forth the interesting suggestion that the decision to retire is made by the officer's wife more frequently than by her husband.

One of the major concerns in the popular press during the post-Vietnam years was the impact of military service on the young men who fought for their country and were then turned back to civilian society alienated and disoriented. M. Kent Jennings and Gregory B. Markus present data from a sample of youth who were surveyed in their senior year in high school, in 1965 and again in 1973. For those who served on active military duty in the intervening years, their involvement in the political system was not, on the whole, negatively affected by their military activity. David R. Segal and Mady Wechsler Segal present data on the impact of service collected from a cross-section sample of the Detroit civilian population, and they similarly find that military service did not have negative consequences, by and large.

This last issue is of special consequence for the all-volunteer military, since every veteran is, in a real sense, a part of the recruiting system. Were military service shown to have negative consequences for the people who served, this would have implications for the public image of the military and the future availability of personnel resources. As John D. Blair and Jerald G. Bachman demonstrate, the public image of the military is not negative—again contrary to some of the views of the communications media. This support is not unqualified. There is evidence of public resistance to high levels of defense spending, and attitudes toward the military have been more negative among young college graduates than among other groups in society.

The attitudes associated with military service in an all-volunteer context emerged as another area of concern. In another paper, Bachman and Blair discuss the possibility of the evolution of a "military mind," not representative of the wider American population. They conclude that career-oriented military personnel do have attitudes discrepant from civilian views and that there are likely to be more career-oriented personnel in the all-volunteer military than was previously the case. Ronald Inglehart provides data which confirms the Blair and Bachman finding that the view of the military among civilians is positive, but which indicates a potential shift in public attitudes involving a decline in support for the military institution. Should the trends identified by Bachman and Blair and by Inglehart continue within the military and civilian populations, respectively, they might well lead to ideological divergence between these groups.

The papers and discussions of the Chicago conference did not produce empirical resolution of the issues raised. They did, we believe, identify a set of important problems which social scientists concerned with the all-volunteer military must address.

The editors of this volume wish to express their appreciation to Professor Sam C. Sarkesian of Loyola University of Chicago, who served as director of this effort and coordinator of the conference, and to thank J.E. Uhlaner and E. Ralph Dusek of the U.S. Army Research Institute for their support, assistance, and advice. Professor Morris Janowitz of the University of Chicago initially suggested the need for such a conference and was a source of guidance throughout. The Chicago conference, as well as the publication of this volume, has been due in large measure to the support and cooperation we have received from the U.S. Army Research Institute for the

Behavioral and Social Sciences and from Loyola University of Chicago.

–Nancy L. Goldman
Chicago, Illinois

–David R. Segal
College Park, Maryland

I.

SOCIALIZATION INTO THE ALL-VOLUNTEER FORCE

THE IMPACT OF BASIC COMBAT TRAINING: THE ROLE OF THE DRILL SERGEANT

John H. Faris

In January 1974 a high-ranking officer at an Army basic training center recounted the following incident in an interview: A graduation ceremony, normally held on the parade field, was shifted to an auditorium. When a noncommissioned officer was presented with the "Drill Sergeant of the Training Cycle" award, the graduating company cheered and applauded with great enthusiasm. The officer expressed much surprise at this apparently genuine and spontaneous display of positive affect. (It is supposed that the indoor setting, being more anonymous and less formal than the parade ground, was less inhibiting.) The fact that the officer was surprised is interesting in itself. Strong feelings of affection for their drill sergeants are very common in trainees, particularly toward the end of training. In a sample of 107 trainees interviewed over a 1½-year period, 91% responded to the question, "What do you think of your drill sergeant?" in a positive way. (None were unqualifiedly negative; the rest were mixed.) Comments such as "He's nice," "He's a great man," and "I love him, oh boy, do I love my drill sergeant" were frequent.

The officer's surprise is understandable. Drill sergeants are thought to be, and are, the source of much unpleasantness for recruits. What is comparably surprising is that not only do most trainees come to like their drill sergeant, but most trainees also have similar feelings about the Army at a time when hardships of Army life are at their peak and privileges are fewest. When asked, in the final week of training, "What do you think of the Army?" 72% responded positively and 11% negatively (remainder mixed). Comments such as "It's really together" and "I love it" were frequent. The findings that most trainees like their drill sergeant and the Army are important in light of the nature of basic training and the criticality of manpower considerations in the all-volunteer situation. This paper suggests that

AUTHOR'S NOTE: *Much of this paper is based on the author's own research-participant observation of basic and advanced infantry training in 1969 and extensive interviews of trainees and drill sergeants and observation of training in 1972, 1973, and 1974 at three different training centers.*

the explanation of the second finding (trainees' orientation to the Army) can be found in the explanation of the first (their feelings toward the drill sergeant) and that an examination of these findings can shed light on the nature of socialization into the all-volunteer ground forces.

Socialization in Basic Training

Although basic combat training is an extraordinary experience, it has been an integral part of our culture. Most men of military age have a father or brother or uncle who has experienced basic training, and dramatic depictions occur with some frequency in movies and television, along with some news documentaries. Most trainees, whether conscripts or volunteers, approach the rite of passage with trepidation. There are several features of basic training which make it extraordinary and which have persisted through the years and appear in much the same form from one post to another.

First, there is at least initially a disparagement of civilian status, which takes the form of degradation and humiliation on both the group and individual levels. Soldiers who have been to college are ridiculed as stupid for not knowing how to fasten an Army belt buckle; an entire platoon is made to feel clumsy for not being able to march as a unit. Unflattering haircuts and glaringly new, ill-fitting uniforms reduce personal dignity. The trainee's fear of authority and his ignorance of what is and what is not a legitimate order (one trainee with a high, thin voice was ordered to chew tobacco) are imposed upon to make the novice look and feel silly.

Second, basic training is characterized by extreme isolation from civilian society on the one hand and an almost complete lack of privacy from other trainees on the other. Contact with friends and relatives is much reduced, while at the same time it is almost impossible to be alone. In many basic-training barracks there are no partitions between the toilets. This is a feature of the experience with which many trainees have difficulty.

Third, much of the evaluation of performance in basic training is done at the group level rather than on the individual level. This collective evaluation (or "group punishment"–which neglects positive reinforcement on a unit basis) violates the trainee's sense of justice. For example, an entire platoon may have passes withheld because the floor under one bunk—the responsibility of one individual—was inadequately swept. This system is the source of many of the strongest complaints (though a minority of trainees perceive the function of such an approach—to develop teamwork and solidarity).

Fourth, basic combat training includes an emphasis on masculinity and aggressiveness. Expressed attitudes toward women are utilitarian and unromantic and tend to reinforce a sense of male superiority. Closely related is a high value on aggressiveness (within the appropriate context—fighting, for example, is rarely tolerated). Trainees are required to growl and scream in a savage manner, to roar "kill" (the "spirit of the bayonet") at bayonet training, and to shout rather than speak in reply to a drill sergeant. The emphasis on masculine toughness combined with the threat of being labeled feminine (in a very derogatory fashion) is traumatic for insecure trainees.

Finally, basic training is designed to place the trainee under various forms of stress, both physical and psychological. While some trainees are in better physical condition than others, mechanisms exist so that almost all experience stress. The stronger trainees may be required to carry an extra 25 pounds of machine gun or radio on a speed march. Trainees doing calisthenics are not allowed to look at one another so that they can locate a group norm; therefore, each trainee may be required to do his own personal maximum of pushups. There are other forms of physical stress—hunger, thirst (in field training), and sleep deprivation. (Sex deprivation is not a particularly frequent complaint. It is plausible that other forms of stress reduce the sex drive, which may account for the persistent rumor that the mess hall food contains saltpeter.)

Psychological stress has a number of sources. Fear of failure and the companion fear of being recycled (repeating part of basic training in another company) are among the most severe types of psychological stress, especially for marginal soldiers. Psychological stress is also generated intentionally by arbitrary and sometimes conflicting demands. One drill sergeant said in an interview that he would see to it that his platoon, which had been doing very well, "won't be able to do anything right tomorrow."

The above characteristics would seem to make basic combat training a highly negative experience, and certainly they are often perceived as negative by the trainees. How, then, can we account for the finding that most trainees respond positively both to the institution which has provided them with this experience and to the drill sergeants who in most cases are the direct perpetrators of these negative experiences? The remainder of this paper will deal with how the drill sergeant manages this transformation.

Characteristics of the Drill Sergeant's Role

The career drill sergeant is a highly skilled manager of small groups. (See Bittner, 1967, for an analysis of similar skills of uniformed police.) A drill sergeant said in one interview:

> Say I have a man with a bad attitude, doesn't care about soldiering. I put him on barracks detail for a few days, just keep him apart, and soon he's begging me to let him return to training. It's like this. Do you like parades? No, of course not. Nobody likes to go on parade. But how would you feel if someone told you that you are the only one who *can't* go on parade, how'd you feel? See?

Some drill sergeants are better able to articulate their knowledge than others, but almost all possess a variety of important skills and attributes which either are acquired through long experience and the inherited tradition or are possessed as a criterion for selection to this duty. These are as follows:

(1) *The drill sergeant as role model.* In a situation where military values are preeminent and the status and achievement of the new trainee are minimal, the drill sergeant presents the image of the ideal soldier. In dramatic contrast to the trainee, the drill sergeant is competent (he has mastered all the skills required of the trainee), assured, tough, ascetic, and attractive—his uniform is usually immaculate, from well-brushed campaign hat to glistening black boots. He is obviously a military leader and is likely to wear the badge of the combat veteran. If the trainee accepts the values of the military (and most do, at least temporarily and provisionally), there is no more appropriate or accessible standard for the enlisted man to aspire to than the drill sergeant. Not surprisingly then, quite a few trainees (36 out of 144, or 25%, in a sample drawn in 1968) form the ambition of someday becoming a drill sergeant. (Almost certainly a major factor in the disenchantment with the Army that frequently

occurs after basic training is the disparity between the quality of leadership provided by drill sergeants and that supplied by NCOs and officers in the rest of the Army. But it would be an error to attribute this disenchantment solely to leadership.)

(2) *The drill sergeant as a father figure.* Many of the senior drill sergeants are in fact old enough to be the trainees' father. Drill sergeants often take a very paternalistic orientation toward trainees, particularly when dealing with them on an individual basis–helping with learning a skill or counseling a soldier with problems. Not infrequently the drill sergeant uses "son" instead of "you," "trainee," "shithead," and other appellations common in other situations. Trainees often come to view their drill sergeant as a substitute father and almost invariably prefer the older, experienced drill sergeant to his younger assistant (who is often virtually the same age as the trainees).

(3) *The drill sergeant as dedicated and caring.* Drill sergeants put in long hours, often as long as the trainees–from 5 A.M. or so to 9 P.M. or, when there is night training, midnight or later. (As a consequence, drill sergeants tend to have unstable marriages.) While, obviously, drill sergeants are not deprived as trainees are, they often do share hardships–they participate in the speed marches, sleep in the woods on bivouacs, etc. Drill sergeants frequently go out of their way to help a trainee with a problem. A captain told of one of his drill sergeants who for more than 3 hours on his day off worked on the family car of a trainee who was being visited.

(4) *The drill sergeant's situational role shifts.* The drill sergeant's orientation in the formal situation of the parade field or company-level training is markedly different from what it is in relatively less formal platoon training, meetings inside the barracks, and individual counseling. As a result, most trainees think that their platoon's drill sergeant is the best in the company, since they have never seen the other drill sergeants in the less formal situations. Similarly, drill sergeants are much more formal and autocratic in "garrison life" and in routine training than they are in field training, particularly in overnight bivouacs and tactical maneuvers. Their manner in the latter situations suggests that this is what the Army is really about, and this impression and the accompanying sense of comradeship are reinforced by the drill sergeant's leaving behind his campaign hat (his peculiarly distinctive symbol) and wearing a steel helmet and pack like the trainees. One otherwise very unapproachable drill sergeant

allowed himself to be called "Sarge" in the field. When the company returned to barracks he reverted to a despot.

(5) *The drill sergeant's role-shifts over time.* As trainees begin to leave behind their civilian attachments, develop their own solidarity, master the soldier's skills, and adopt the values of the military, the drill sergeant comes to treat them, not as equals, but as worthy of eventual acceptance into the status of soldiers. There is increased relaxation and approval. One of the elements of this process, the use of humor, is important enough to be examined in depth.

Most drill sergeants use humor with great effectiveness. The first form in which this appears is ridicule, part of the degradation process. At this stage, when the trainees as a collectivity (not yet a group) are the target, it is not intended or perceived as humorous, though the form is humor. When an individual trainee is the target, the other trainees are often not permitted to engage in laughter or join in the ridicule. They are not allowed to imply that they are in a different category from the unfortunate victim.

> The first time our drill sergeant met the platoon was the morning after our haircuts. We stood rigidly in what we thought was the position of attention. He stalked silently through the ranks, examining us for a few minutes. Finally, "Where's my two hippies?" Silence. "All right now, Goddammit, where's my hippies? They told me I had two longhairs." After a long and uncomfortable silence Private W. said, "I had long hair, sir." "Sir? Sir! Don't call me sir, Goddammit, I work for a living. What are you, a sissy? What d'you wanta wear that bebop hair for?" And so on, inches away from W.'s face, amidst our silent gratitude and pity.

The effectiveness of ridicule is in part dependent on the nature of the platoon at this stage. Trainees are almost complete strangers to one another, although within hours most begin to develop friendships with a bunkmate or two. The group is atomized and offers little defense to the assault of initial socialization.

Before many days, though, a wider and deeper solidarity emerges. There is a development of trainee humor both as a defense against the drill sergeants and basic training and as a means of reinforcing solidarity. A jocular griping, often coarse in form, is widespread in basic training. Some standard gripes become institutionalized —creamed beef on toast is known universally as "shit on a shingle."

Another example of the cohesion-producing function of humor is provided by Obrdlik (1943). His discussion of the gallows humor of

Nazi-occupied Czechoslovakia emphasizes the importance of this humor for social control and morale.

> These people have to persuade themselves as well as others that their present suffering is only temporary, that it will soon be all over, that once again they will live as they used to live before they were crushed.

Obrdlik also points out that gallows humor can be aggressive and can undermine the morale of oppressors. In basic training this can threaten the position of the drill sergeant. An inexperienced and insecure officer is particularly vulnerable, as this account of a young officer with his first command shows (Fraser, 1970:26). The officer had sympathetically rescinded the formal reprimand of a soldier whose rifle was dirty and then had overheard that same man laughing in the barracks.

> I moved on a few steps. All right, he had made a quick recovery. But was relieved. There was nothing in that. But he had seemed really upset in the armoury, shaken, as Cormack said. Now he was roaring his head off—the quality of laughter somehow caught the edge of my nerves. I stood undecided, and then wheeled around and shouted:
>
> "Sergeant Telfer!"
>
> He came out of his room. "Yessir?"
>
> "Sergeant Telfer," I said, "stop that man laughing."
>
> He gaped at me. "Laughing, sir?"
>
> "Yes, laughing. Tell him to stop it—now."
>
> "But . . ." he looked bewildered. "But . . . he's just laughin' sir . . ."
>
> "I know he's just laughing. He's braying his bloody head off. Tell him to stop it."
>
> "Right, sir." He obviously thought the sun had got me, but he strode into the barrackroom. Abruptly, Leishman's laughter stopped, then there was what might have been a smothered chuckle, then silence.
>
> Feeling suicidal, I went back to my billet. Obviously, Leishman had thought I was a mug; I should have let the charge stick. Let someone get away with it, even a good soldier, and you have taken some of his virtue away. On the other hand, maybe he had been laughing about something else entirely; in that case, I had been an idiot to give Sergeant Telfer that ridiculous order. Either way, I looked a fool. And my Service dress didn't fit. To hell with it. I would see the adjutant tomorrow and ask for a posting.

But even the experienced trainer is not immune to an extraordinary assault, as indicated by this account by T.E. Lawrence (1955:117) of his basic training experience in the RAF under the assumed name of Ross. A fellow recruit, Sailor, comments to Lawrence:

> "But, mate, you let the flight down, when he takes the mike out of you every time. Give the ignorant shit-bag a ––– great gob of your toffology."
>
> Next day, in the first stand-at-ease of first period:–"Short-arse you there, Ross, what's your bleedin' monaker:–what d'you know?" Such non-plussing questions are Poulton's favorite gambit for a hazing. *Spring to attention.* "Sergeant," I dutifully intone. He wouldn't stop. "I asked you a question, you little –––." But I am not tired at this time of day: by Sailor's advice of priggery I made to drawl out "Well, Sergeant, specifically of course we can know nothing–unqualified–but like the rest of us, I've fenced my life with a scaffolding of more or less speculative hypotheses."
>
> The rear rank deflated appreciatively. The tired sounding, like the wind in wet trees. The Sergeant stared; then whispered to himself "Jesus ––– Christ." At that Sailor let out a high, sudden, singing laugh. "Flight. Attention" Poulton yelled, and the drill went forward, gingerly. "My Christ" exulted James, thumping my back later, in the hut's safety, "The silly ––– didn't know whether his ––– was bored, punched, drilled or countersunk."

Drill sergeants anticipate, preempt, and transform this process by gradually shifting from ridicule to a joking relationship. There emerges, too, the use of banter. Thus the drill sergeant manages two statuses–the superordinate noncommissioned officer socializing agent and the prospective fellow soldier and comrade-at-arms.

Banter may be verbal or may exist in the ludicrous character of a situation.

> One of the men in my squad was caught with a candy bar in his laundry bag, strictly forbidden. The drill sergeant inquired, "Who's your squad leader?" He then took the candy bar, mashed it with his heel in the dirt, and directed me to put it in my pants pocket and keep it there for the rest of the day. It was a hot summer day.
>
> One day, returning on a bus from successful training, M., a black trainee, asked the drill sergeant when he could come over to his house for dinner. The drill sergeant said that he was from Tennessee and that M. didn't have enough money to get through his front door. M. said, "C'mon Drill Sergeant, I got enough money to burn a wet elephant." All of this was in high good humor, much appreciated by the rest of the platoon.

The essential element of this joking relationship is that it requires the cooperation of the trainees. Daniels and Daniels (1964) have indicated the importance of the role of the good-natured fool in providing the focal point for the joking relationship between the drill sergeant and the trainee group.

The function of the fool as a social type makes a dramatic contrast to the sick role in basic training as described by Schneider (1953:396).

> The sick role develops first as a group-sanctioned behavior, focusing the resentment of the group against the army, providing a *cause célèbre* for aggression against the army. But the group sanction lasts only until the group has begun to make its adjustment to aborted affectional needs, heightened masculinity demands, and brother-like relationships. When the group begins to make its adjustment, its need for rationalization of its aggression against the army diminishes, so that it no longer supports the sick in their role. At this time the divorce between the sick and the group takes place, and the sick find themselves isolated from the group.

This provides the limiting case to the proposition of Dentler and Erikson (1959) that cohesion forms around deviance.

The joking relationship that develops between trainees and drill sergeant is constrained by bounds. It is appropriate in some situations (riding on a bus) and unthinkable in others (inspection in ranks). Some topics, being sacred, are not permissible for joking —e.g., the drill sergeant's dignity:

> After joking about dinner at the drill sergeant's house, M. made some joking remark about the drill sergeant's Smokey-the-Bear hat. This was greeted with great laughter by the platoon and provoked a threat by the drill sergeant to put us all to cleaning the mess hall grease trap with a spoon. The rest of the bus trip was in nervous silence.

There are sacred topics for trainees as well. A drill sergeant's question about the sex life of a married trainee met such stony resistance that the subject was dropped.

The skillful use of humor as a socialization mechanism is an important factor in effecting the transformation from a cohesive group with oppositional tendencies to an accommodation with and adoption of the values of the institution, while retaining the distinctions between the status of the private recruit and the veteran NCO. It is a form of seduction.

(6) *Drill sergeants as perpetrators of an illusion.* Drill sergeants perform a sort of magic act which manages a serious contradiction —that between the threat of failure and the challenge of basic training, on the one hand, and the reality that almost everyone succeeds, on the other. While many trainees fear being recycled, only a few are (1-3%) and usually as a result of serious injury or illness. Failure on the comprehensive test at the end of basic training (especially since the test was made more rigorous in 1972) is a very real possibility; often fewer than 20% pass on the first administration. But almost everyone passes the second time, and virtually all the rest are "pushed through" even if it takes one or two more tries. This situation results in role strain for the drill sergeant. The organizational demand for high success rates in a situation of scarce manpower conflicts with his professional norms—that only a well-trained soldier should be permitted to graduate. Drill sergeants complain that they have to cajole, and even bribe, examiners to let the hopeless cases through. But this is the extreme case. In any event, drill sergeants must create the perception that basic training is an extraordinary challenge, one which will set the graudates apart from others, in face of the fact that almost everyone succeeds. The fact that most trainees begin to discover this is indicated by the frequent comments toward the end of training that they wish it had been tougher. When pressed further, though, they usually admit that at the time it seemed hard enough.

Transition to the All-Volunteer Army

These techniques enable the drill sergeant to perform a striking transformation—to generate a strong articulation with institutional authority and values in a situation which has an inherent tendency to create oppositional solidarity (e.g., the prison inmate subculture). As Shils and Janowitz (1948) pointed out, this articulation is as necessary for combat effectiveness as the existence of the primary group.

During the transition to the all-volunteer Army (approximately 1971-1973), several factors emerged to jeopardize this process of socialization. First, many recruits entered the Army with unrealistic expectations of a transformed basic training, developed through a large advertising campaign and, more important, the sales pitch of recruiters struggling to meet quotas. For example, 8 hours of sleep per night was guaranteed but rarely delivered. (Even during this period, most trainees expected basic training to be harder than they found it; the generalized conceptions are too powerful to be quickly changed.)

Second, drill sergeants were confronted with new restrictions on their behavior toward trainees. They were forbidden to curse, ridicule, or, in some instances, even raise their voices to trainees. These restrictions were widely ignored, but the contradiction between policy and practice was apparent. In 1972 one drill sergeant advised an interviewer to go to the commanding general's orientation for new trainees and see how the general's remarks on the first day destroyed the purpose of basic training. The general, among other things, promised trainees that they would not be degraded, that as long as they acted like men they would be treated like men.

Third, the mechanism of collective evaluation was threatened by an increasingly individualized approach to basic training, especially in a new self-paced style of weapons training, first aid, and so on. However, in practice, the new procedures were transformed into traditional structure; trainees who were supposed to have mastered a skill (demonstrated by a test) continued to practice it along with the others instead of moving on to new material.

Fourth, the impact of basic training was reduced by shortening it by 2-3 weeks to less than 6 weeks. Concurrent with this, there was a relaxation of pass policies. Contact with civilian ties was much less disrupted than previously.

Fifth, changes in the content of basic training removed some of the important rites of passage. Most significant of these is the infiltration course, which involves crawling under live machine gun fire. Trainees who had heard "war stories" from brothers and friends about this test of their manhood were disappointed to find it no longer part of the program.

By January 1974, many of these characteristics of the transition were being again transformed. Basic training was being lengthened; the content was becoming more traditional (more drill and ceremonies, less advanced weapons training); the infiltration course was reinstituted; and local commanders were returned more discretion over the conduct of training. In the all-volunteer Army, then, the impact of basic training and the role and effectiveness of the drill sergeant in the socialization process are increased by several factors:

(1) Volunteers tend to bring to basic training a favorable orientation toward the Army as a whole, if not toward basic training.

(2) Volunteers are on the average younger than conscripts and often come from homes in which the father is weak or absent. The paternalistic orientation of the drill sergeant is particularly powerful.

(3) Many volunteers are lacking in self-esteem, their lives being patterns of failure and deprived of direction. Their success in basic training at acquiring skills (marksmanship, communications, etc.), meeting challenges, and enduring hardship is the source of much gratification. Going home on pass can have dramatic consequences, for many trainees report perceiving their friends as childish and immature, and their former activities as meaningless. Some trainees home on pass are apparently effective recruiters, several claiming to have convinced friends to enlist (one claimed five).

(4) The end of active ground-combat involvement in Southeast Asia has reduced the ambivalence generated by the socialization process, on the one hand, and the fear of combat in general and the widespread abhorrence of this war in particular, on the other.

The drill sergeant and the organization of basic training perform their function with effectiveness. One of the Army's major adaptations to the all-volunteer situation which has yet to be resolved is to find ways of retaining the peaks of morale and solidarity in the very different situations of advanced training and active duty, where there appears to be a serious letdown in morale and satisfaction (U.S. Department of the Army, 1971; Goodstadt et al., 1975).

II.

THE IMPACT OF MILITARY COMBAT

THE REENTRY TRANSITION OF THE VIETNAM VETERAN

Jonathan F. Borus

My primary research interest over the last several years has been to study the processes of individuals and institutions coping with change. Most of my work has explored transition periods, common times of significant life change, in attempting to understand the process of "normal" people moving from one major phase or stage of their lives to another. Over the last several years my work has successively focused on inner-city and ghetto youth making the transition to college life, on military personnel making the transition from our segregated civilian society to the forced integrated society of the Army (Borus et al., 1972, 1973; Fiman et al., 1975), on Vietnam combat veterans making the transition from the battlefield back to the States (Borus 1973a, 1973b, 1973c, 1974), and on some recent studies of young physicians making the personal-professional transition from the trainee-prolonged adolescence stage to the practitioner-full adult stage of their lives.

In the normal life cycle there are expectable transition phases from one life stage to another and from life situations requiring one set of behaviors to other situations requiring quite different behaviors. Transitions, especially rapid ones, are often stressful periods when past adaptational mechanisms may be neither adequate in nor applicable to the new situation. Especially stressful transitions, called "crises," are times of emotional disequilibrium characterized by the potential for successful adjustment through development of new and appropriate coping methods or maladjustment through inappropriate application of unsuitable coping behaviors (Tyhurst, 1957). In this context, coping can be defined as the "problem solving efforts made by an individual when the demands he faces are highly relevant to his welfare (that is, a situation of considerable jeopardy or promise) and when these demands tax his adaptive resources" (Lazarus et al., 1974:250-251).

In psychiatry, most effort has been devoted to clinical work with individual patients. From this, there frequently comes the retrospective deduction that current pathology may have arisen or been precipitated by inadequate or maladaptive coping with internal or external sources of stress during earlier life transitions (e.g., inadequate grieving at the death of a loved one, social withdrawal and loss of self-esteem following interpersonal rejection, etc.). Rarely have we had either the opportunity or the inclination to study systematically the coping issues and coping methods utilized by nonpatients at the time of life change itself (the "transitional synapse"). I belive that if social psychiatry is to become relevant, it must study the common periods of life transition in order to define the salient adjustment issues, distinguish what makes for successful as opposed to unsuccessful adjustment, and design interventions to foster successful coping.

It may be more than coincidental that the word "cope" itself has a martial derivation, coming from the old French verb *couper* ("to strike"); it denotes a process of struggle, as in warfare, when the individual must mobilize his resources and allocate them to offensive, defensive, and systems-management operations to ameliorate significant stress. The entire experience of military service is characterized by a multitude of often stressful transitions with which the soldier must struggle and which demand frequent utilization of coping methods to adapt or adjust to new environments; in fact, change is one of the few certainties of military life. Consider briefly the string of transitions in status, role, and location (most not within the control of the individual soldier) of the average military career: from the civilian environment and role to basic combat training to advanced individual training in another location to a series of post assignments (which always include the possibility of being in a foreign, quite possibly hostile, and perhaps dangerous combat environment) to the return after some period of time to a new or renewed civilian life and role. During all these stages, the soldier must cope with certain pervasive changes which differentiate military from civilian life: the forced integrated setting of the military, changes in his basic constitutional and civil rights, a decrease in his ability to control his life (with a much greater dependency on others), and the generalized demand for his adjustment to a variety of both extreme and mundane environments.

How a soldier copes with these multiple transitions during his military career depends on his past history of coping, the flexibility and variety of the coping repertoire he has established in adjusting to past transitions, the institutional supports or lack of them provided him by the military, and the availability of immediate group support from significant others who are undergoing or have undergone similar transitions. The transition points in military careers are often times of difficulty for the individual soldier and of loss of effective manpower for the military unit; e.g., a sizable portion of all AWOLs occur immediately prior to, during, or immediately after a soldier changes assignments in making his transition from one part of his military career to the other. There is a great need for the military and its behavioral scientists to delineate formally the expectable transitions in military careers and to design implementable programs applicable to large groups of soldiers to facilitate maximally effective and individually integratable transitions in location, status, and role. In addition, it would be helpful to attempt to plan for some of the more idiosyncratic stresses, which can be predicted to occur frequently in populations as large as military ones, so that programs of support for idiosyncratic change are also available to individual soldiers.

Two factors appear to interfere with rational planning by the military around facilitating healthy coping and adjustment during transitions. The first is a fear that I have heard expressed by high-ranking military commanders which might be called "the self-fulfilling prophecy apprehension"; that is, if the military focuses on its transitions as possibly disruptive, acknowledges that change occurs, and plans and implements helpful interventions, this very activity may produce an expectational set for soldiers that they *should* have difficulty with transition periods—thus *causing* an increase in the number of people who do have difficulty. This leads some commanders to assume an ostrichlike posture of ignoring potentially disruptive and stressful situations rather than figuring out ways to facilitate soldiers' coping with them. An example of this from my experience was the high-ranking commanders who were apprehensive about ascertaining the nature of the racial climate in their units through use of the Racial Perceptions Inventory for fear that asking about race relations would *cause* racial strife. They acknowledged their hope that, if only they ignored race relations in their units long enough, they personally would enter their next military transition (move on to command another unit) before racial disruptions occurred for which they would be held liable. The other factor inhibiting such planning seems to be the fact that, percentage-wise, the majority of healthy young American adults in the military are able to adjust successfully to each of the military transitions without significant help; this leads to the projective tendency, especially in peacetime, to designate those who do experience difficulties with the transitions as idiosyncratically deviant and therefore easily dispensable through discharge back into civilian life. This stems from the military's primary mandate to provide an effective fighting force rather than healthy individual fighters and from the lack of funding to the military to provide remedial and rehabilitative, let alone preventive, services for soldiers to facilitate coping with the stresses of noncombat military life. The military's most formal and rigorous preparation for change, basic combat training, prepares soldiers for combat roles which most of them never perform; most military roles are maintenance rather than martial in nature. We prepare soldiers fairly well for such extreme environments; we do almost nothing to help them cope with the more usual mundane environments of the noncombat military with which many have significant difficulties. In fact, the military has been better able to plan for the truly idiosyncratic event—e.g., how to get a soldier home on emergency leave during the terminal illness of a parent or family member—than for the expectable stressful situations undergone by a large number of its personnel, such as preparing soldiers in advance for the changes to be expected in their reassignment to a new noncombat post and role.

Studies of the Reentry Readjustment of Vietnam Veterans

My interest in the coping processes of Vietnam combat veterans began in 1969 when I was involved in a personal transition from civilian to soldier. While doing my basic officer's training at the Medical Field Service School at Fort Sam Houston, I became aware of the significant life changes that I was undergoing in making this transition and of the even greater and more stressful demands on my cohorts who were to be assigned to the Vietnam combat zone immediately after basic training. I saw our immersion in a course which basically was preparing physicians for the combat zone (preparation in tropical medicine, methods of calling in a medevac helicopter, crawling through an infiltration course under live fire, learning international law protecting prisoners of war, and learning methods of possible escape from imprisonment) as helpful in stimulating an adjustment of the expectational sets of my physician friends who were going directly to Vietnam; for some of them it added a spark of anticipatory readiness and almost eagerness about their upcoming service in the combat zone to their undeniable concurrent sadness and depression at leaving behind family, friends, and familiar environments and roles. This experience also made me question how it was preparing me and the majority of physicians who would not go to Vietnam for noncombat military service; although I probably would not have to face a battlefield situation similar to the infiltration course, I would have to cope with the large military and military medical bureaucracies over the next 3 years. While being favorably impressed with the preparation for the combat zone, I also began to wonder whether there were similar preparation and support for people returning from the combat zone to the more mundane but also stressful environments of stateside military and civilian life.

I spent some of the introductory days of my assignment at the Walter Reed Army Institute of Research getting to know the stateside Army routine at a garrison post in nearby suburban Maryland, where I was introduced to a number of the care-giving agencies available on garrison (mental hygiene, Army Community Services, military justice system). Representatives of these care-givers spontaneously expressed their concern with the difficulties experienced by the number of troops at the post who had returned from Vietnam, and, much to my surprise, I found that there were no formal or informal programs to facilitate their readjustment from the combat to the noncombat military. Over the next several years I found that this was a general phenomenon and that specific programs aimed at helping Vietnam veterans readjust simply did not exist in the Army.

On reviewing the literature of American veterans returning from combat, specifically the major works of World War II (Stouffer et al., 1949; Grinker and Spiegel, 1945), I saw clearly that, in the past, military behavioral science planning efforts and resources had been utilized for interventions to deal with the expectable stresses of the transition back to the States. From the World War II experience, the following were found to facilitate the transition: the helpfulness of a gradual rather than precipitous transition from combat to noncombat status; the need to reorient combat soldiers to the very different lives and roles that they would occupy in the stateside military; the need for formal and ceremonious acknowledgment to the soldier and his significant others of his change in status and the expectancy of the need for different coping responses to new stresses; the helpfulness of the immediate group or small unit in sharing experiences and facilitating readjustment; the need to prepare the veteran in advance for the adjustment stresses he might have in assuming his new military and civilian roles; and the need to make the continuing noncombat military role meaningful to the veteran. These readjustment areas were explored, attempted, to some extent evaluated, and widely supported in World War II, since there were genuine concern and action to facilitate successful coping by veterans. There was little written about veterans returning from Korea, perhaps because that was an unpopular and not totally successful endeavor to which we were even loath to give the formal appellation of "war." Similarly, when I began my studies in 1969, almost nothing had been written about Vietnam veterans. They were essentially ignored by the military for the first 8 years of the war, until finally congressional and public pressure in the early 1970s, followed by the political and military consequences of drug use in Vietnam, brought their plight into the professional and public spotlight (U.S. Senate, 1970, 1971).

To attempt to shed some light on this ignored area, in 1969-1972 I conducted two concurrent studies to attempt to understand both the process and the outcome of Vietnam veterans' attempts to adjust within their initial period of reentry, which I arbitrarily designated as their first 7 months after leaving the combat zone. Most Vietnam veterans spent this reentry period completing their active-duty military obligations at a stateside garrison post. The first study was a prospective cohort investigation of the incidence of reported maladjustment in Vietnam returnees during their first 6 months back at a stateside garrison unit (first 7 months after leaving the combat zone) in an attempt to delineate the percentage of this veteran population experiencing difficulties in readjustment as compared to that of a control group of non-Vietnam veteran soldiers in the same unit (Borus, 1974). The second was an interview study to determine and understand what are the stressful issues with which returning Vietnam veterans must cope, what differentiates those who are successful in coping with these issues from those who have difficulties, and what programmatic interventions may appropriately be implemented to foster and facilitate healthy readjustment on return from Vietnam (Borus, 1973a, 1973b, 1973c).

METHODS

The entire first-tour enlisted-rank population (N = 1009) entering a large combat unit of 3,000 men at an East Coast garrison post during the last 6 months of 1970 were identified and studied. Of these new arrivals, 765 were Vietnam veterans who had just completed a year's tour in Vietnam (primarily in combat roles as infantrymen or artillerymen) and had reported to the garrison unit immediately after their one-month post-Vietnam leave. The remaining 244 nonveteran soldiers came to the unit either from noncombat overseas duty elsewhere (e.g., in Germany or Korea) or from noncombat units within the United States. Post, regiment, and individual military service records were monitored for each soldier for a 6-month period following his arrival at the garrison unit and complete data were obtainable for 749 (577 Vietnam veteran, 172 nonveteran) subjects. The most available measures for quantitative study of military records were indices of disciplinary-legal maladjustment (individual or unit records indicating punishment for an infraction of the Uniform Code of Military Justice) and emotional maladjustment (records indicating utilization of the regimental or post mental hygiene facilities). The third outcome group, that of satisfactory adjustment, was assigned to an individual who had neither received disciplinary-legal punishment nor utilized mental hygiene services for emotional difficulties during the entire follow-up period. Incidence rates of disciplinary-legal maladjustment and emotional maladjustment, adjusted for person-months at risk, were calculated. These rates for the veteran and nonveteran groups were then compared, using the critical ratio test for association.

As the monitoring of the incidence study proceeded, subjects for the interview study were identified. Soldiers whose records indicated that they were experiencing either disciplinary-legal maladjustment or emotional maladjustment were then interviewed by me in a 2-hour semistructured interview under conditions designed to guarantee anonymity to the interviewee. In an attempt to control for socioenvironmental variables, I also interviewed veterans who were satisfactorily adjusting and who lived in the same military company, during the same period of time, as one of the maladjusting interviewees. The total interview sample of 64 veterans comprised three approximately equal-sized groups: adjusting (Group A, N = 22), legally maladjusting (Group B, N = 22), and emotionally maladjusting (Group C, N = 20).

RESULTS OF INCIDENCE STUDY

The results of the incidence study showed that only 23% of the Vietnam veterans had some record of maladjustment in their first 7 months back in the United States. A total of 21.5% of the veterans had been in disciplinary-legal difficulty and 4.1% had encountered emotional difficulty since returning from Vietnam (these percentages included 2.6% of the returnees who had indices of both disciplinary-legal and emotional problems). Of the veterans with indices of adjustment difficulties, 63% had only one disciplinary-legal offense during the follow-up period, and only 1.1% had adjustment difficulty severe enough to warrant premature discharge (psychiatric, general, or dishonorable) for maladjustment reasons. Surprisingly, when the indices of maladjustment for Vietnam veteran subjects were compared statistically with those of the nonveteran control group, there were no significant differences between the groups. In the control group of nonveteran subjects, 20.3% had been in disciplinary-legal difficulty and 9.9% had encountered emotional difficulties (including 4.1% with both disciplinary-legal and emotional problems) during the period of study. These results suggested that, in contrast to subjective reports of widespread severe post-Vietnam readjustment difficulty, a relatively small (but significant) minority of the veterans in this active-duty population had recorded indices of maladjustment during their initial reentry transition back to life in the States. That the incidences of maladjustment for Vietnam veterans were not greater than those of nonveteran controls in the same social milieu challenged the assumptions that the Vietnam experience or the reentry transition itself was a debilitating stress for the majority of returning veterans, at least during the initial 7 months of their return. The variance of these results with many of the other published reports on veteran readjustment may well stem from different time frames of study, utilization of different indexes of adjustment-maladjustment, the controlled nature of this study, and its focus on an unselected, active-duty sample of veterans in contrast to the samples of many other reported studies which have focused on psychiatric patients, antiwar veterans, and/or veterans with acknowledged readjustment difficulties (Goldsmith and Cretekos, 1969; Strange and Brown, 1970; Braatz et al., 1971; Solomon et al., 1971; Shatan, 1973; Lifton, 1973a; Gault, 1971).

RESULTS OF INTERVIEW STUDY

The incidence study did, however, point out that a sizable number of these veterans who had successfully navigated the combat zone had some level of difficulty readjusting to the stateside military. The concurrent interview studies defined the common adjustment issues which faced both adjusting and maladjusting returnees, described the differences between those veterans who were successfully coping on return and those who were not, abstracted differential coping methods utilized by each study group in dealing with the readjustment transition, and proposed a preventive intervention program to facilitate healthy coping in the reentry transition.

THE ADJUSTMENT ISSUES

The adjustment issues of the reentry transition included those of military adjustment to the noncombat Army life of the garrison post. The veteran was confronted with the major task of reworking his affiliations to both his unit and the Army as an institution, in large part because of the great differences between Vietnam and garrison mission, group support, and leadership. In contrast to Vietnam, these combat-prepared returnees felt that their garrison units had no useful or productive mission for them to carry out and that their time was being wasted in the Army by performing meaningless and degrading jobs unrelated to their training. They missed the closeness and group support of the combat unit and felt that their present unit peers were neither supportive to their commitment to the Army nor demanding of their affiliation to the unit. They felt little confidence in the leadership at the garrison post, as contrasted to reports of great confidence in their leadership in Vietnam; and, on the whole, they felt it was difficult to adjust to the more formal military discipline encountered on returning to the garrison post. They said they had received *no reorientation* to the new demands of garrison life to help them cope with the military aspects of the reentry transition.

Issues of family adjustment centered on the stress of the returnee's initial reunion with his family on arriving home from Vietnam, his adjustment to the often changed and disrupted family situation during his 30-day leave, and his attempts to reintegrate into the family on a part-time (weekend) basis after return to active duty at the garrison unit. In each of these phases, the discrepancy between the veteran's well-nurtured fantasies about home and the reality of his return to his family had to be resolved. In essence, many of the soldiers were destined to be disillusioned about their home and family because they had nourished hypertrophied illusions of them during their year in Vietnam. For many, also, the relative deprivation of being near to home and yet still unable to partake fully in family life because of continuing military obligations was more difficult to bear and more frustrating than the absolute deprivation of family life when they were thousands of miles away in Vietnam.

Issues of social adjustment faced the returnee with coming to grips with his participation in Vietnam in light of the society's antiwar, antimilitary posture, with defining his position on the war and the antiwar movement, and with readjusting to the change in racial relations in the garrison. Most men had well-defined pro- or antiwar positions about Vietnam, denied feeling sorry or guilty about their participation in the war, and did not feel slighted by society upon return. After a loosening of racial stereotypes and close interracial relationships in combat units in Vietnam, veterans reported racial polarization and a reconcretization of racial stereotypes on return.

Emotional issues faced by most veterans included working through changes in emotional temperament related to Vietnam and keeping in touch with and expressing feelings in a socially acceptable, tension-resolving way. Some veterans also faced the more serious adjustment tasks of resolving recurrent emotional disturbances about combat experiences, gaining control of violent feelings and behaviors, and gaining perspective and control over patterns of illegal drug use initiated or exacerbated in Vietnam. Most men felt somehow changed by their Vietnam experiences; the people who were having difficulty on return generally felt changed in a damaged way, while those who were having little difficulty felt the experience either did not change them or, in some way, changed them for the better. Most returnees felt more at ease discussing and sharing their Vietnam experiences and feelings with other Vietnam veterans than with people who had not been to Vietnam. Several veterans mentioned their difficulty in talking about Vietnam to relatives or friends, who would then tell the veteran what it was *really* like from viewing the war on their television screens.

DIFFERENTIAL COPING

The interview studies were also aimed at trying to understand the differences between those soldiers who had difficulty readjusting and those who did not. Demographically, four variables, all related to past and present coping success, retrospectively differentiated the three groups. Compared to the maladjusting soldiers in the other two groups (B and C), the Group A soldiers were significantly better educated, held higher rank, were older, and were more frequently drafted into service. The former two variables indicate Group A's achievements in coping with the stresses of complex educational and military institutions, while the latter two variables indicate their success in finding and holding the job or schooling from which they were drafted at an "advanced" age. Compared to their less successful cohorts, these adjusting returnees in Group A were best able to express their feelings about Vietnam without denial or acting out, had little difficulty keeping conflicting family and Army roles in balance, easily sublimated violent affect, were least likely to be on drugs, and felt the least changed by the Vietnam experience. They used the following variety of coping behaviors, built on past histories of adjustment success, to put the present stresses of return in perspective and make them tolerable. They were able to contain their anxiety sufficiently to consider the long-range consequences of impulsive acts and to plan for the future so that the present stresses seemed minor in terms of long-range plans. They understood the functioning of large institutional systems and learned to deal with the present institution, the Army, in an effective way. They understood the stresses of the garrison post—e.g., the details, the guard duty, the discipline—as an unpleasant part of a bureaucratic system which was not personally directed against them and they knew how to use both formal and informal communication channels appropriately to get what they wanted out of the Army. For example, if a Group A soldier wanted to go on a weekend pass, he knew *both* how to formally request it in a way most likely to get a favorable response and how to informally take it without being marked AWOL if formal permission were not likely. In sum, these soldiers' broad coping repertoires enable them to contain anxiety and maintain self-esteem in the transition period of return while accomplishing tasks to the satisfaction of themselves and the Army.

Group B, the maladjusting soldiers who had disciplinary-legal problems upon return, were intermediate between Group A and C in their variety of coping behaviors and their coping success in working through and expressing their violent feelings about Vietnam, controlling violent behavior, dealing with family problems, and resolving drug habits. They basically contained their anxiety and tried to maintain self-esteem by the coping method of denying readjustment conflict in a defensive, rigid way or expressing internal conflicts through impulsive actions, often relying on outside authorities (family or the Army) to keep them in line. They had little understanding of how large institutions work and were regularly unsuccessful in attempts to get what they wanted from the Army without getting into trouble. These soldiers usually felt harassed by the military stresses of garrison life and did not understand the differentiation and subtleties of the Army's formal and informal communication networks well enough to be able to manipulate them to their own benefit. For example, a Group B soldier invariably got caught if he violated a rule, because he did not understand the system sufficiently to use the proper timing or approach. In sum, these soldiers' coping attempts to contain anxiety and maintain self-esteem emphasized denial of the internal readjustment stresses or expression of them in action, with variable task performance which at least once (the inadequate performance which occasioned the military punishment that defined their inclusion in study Group B) was unsatisfactory to the Army.

Group C soldiers, who experienced emotional difficulties on return, unsuccessfully tried to escape the transition situation by adopting a variety of avoidance behaviors rather than by actively coping with it. They felt the most damaged by the Vietnam experience and had the most difficulty working through and expressing feelings about Vietnam, controlling violent affect and behavior, resolving nightmares and obtrusive thoughts from the Vietnam experience, dealing with family problems, and controlling drug use on return. They tried physically and psychically to escape the stresses of readjustment by withdrawing or avoiding contact with others, or they tried to manipulate the Army into transferring or discharging them. Some, more adaptively, sought professional help on their own initiative in order to try to learn better methods of coping with their problems. In sum, their unsuccessful coping behaviors centered around avoidance and neither contained anxiety nor maintained self-esteem to allow success in task accomplishment.

Preventive Intervention

From these interview data, a preventive intervention program to facilitate healthy readjustment in Vietnam veterans during their initial reentry stage of readjustment was proposed. This program was built upon the principles of acknowledgment of the normality of stress in the transition period of reentry and the need for gradual accommodation to the changes of a new milieu (Caplan, 1971). It emphasized the provision of advanced information about expectable stresses and group support from other combat-veteran peers to facilitate successful coping with such stresses (Hamburg and Adams, 1967). It was suggested that a program to foster successful coping in the transition from the combat zone to the United States should most logically begin before the soldier left the combat zone. This would provide an opportunity to anticipate and rehearse with a relevant peer group the possible stressful situations that the veteran might encounter on leave as he returned home to his family and friends for the first time in a year or longer. It could also provide information about helpful resources near his home so that he would know where to turn should things go wrong during his post-Vietnam leave.

When the veteran was assigned to the garrison post after his leave, the program suggested that he attend a series of rap sessions led by paraprofessional combat-veteran readjustment counselors who themselves had recently gone through the reentry transition. These sessions would serve the purposes of describing the normative adjustment issues and expectable stresses of return; encouraging returnees to rehearse ways to deal with these stresses; allowing the returnees to discuss and work through common feelings and experiences relating to the combat duty in an atmosphere sanctioning the sharing of feelings and experiences with peers as acceptable and beneficial behavior; providing a "basic noncombat training" orientation to the demands, expectations, and organizational structure of garrison life; teaching the returnees methods of working within the military bureaucratic system to achieve self- and institutional satisfaction; providing examples of how others have successfully coped with the reentry transition; and informing the veterans about available helping resources should difficulty in readjustment occur.

It was proposed that such rap sessions would continue on a weekly basis during the first 6 months of return with the backup of mental health professionals who could serve as supervisory resources to the paraprofessionals, as therapists for those returnees with more serious emotional difficulties, and as channels of communication to both unit commanders and staff policy makers with the aim of decreasing some of the unnecessarily stressful aspects of reentry.

It was speculated that such a program would facilitate successful coping with the issues of military readjustment by helping soldiers to anticipate garrison expectations and stresses, teaching them methods of dealing with the frustrations of garrison life, providing a vehicle for greater cohesiveness and support among unit peers, and providing feedback of information to unit leaders and policy makers to help them better understand their men. Such a program would be useful in coping with issues of family adjustment by stimulating more realistic anticipation of the initial reentry into the family and providing ongoing resources for help and peer support in time of transition difficulty. Successful coping with the issues of social readjustment would be facilitated by the program's provision of a vehicle for both the expression and the understanding of feelings about participation in the Vietnam War and continued interdependent black and white verbal interaction. Emotional issue resolution would be facilitated by provision of a sanctioned setting for veterans to talk with peer returnees about upsetting combat experiences, to see that many feelings and reactions were shared by others, and to discuss changes that the veteran and others perceived in him as a result of combat duty. Such self-help groups would encourage verbal expression of affect, disclosure of difficulties, and mutual support during a period of common stress. Last, such a program would facilitate early recognition and rapid professional help for the minority of returnees whose severe emotional and interpersonal difficulties required therapeutic interventions.

I was somewhat disappointed that the intervention ideas generated by my studies were never followed through by the military (to the best of my knowledge) and seemed to have been "circularly filed." Perhaps this occurred because the war no longer directly involved American soldiers. However, the lack of interest and lack of concerted input of resources into facilitating readjustment throughout the 10 years of active American military involvement make me question the commitment of the military to such important endeavors. On the whole, the military appears to me to be trying to forget the war and its possible consequences on the men who fought it. To my knowledge, there are no longitudinal controlled studies by the military or the Veterans Administration of "typical" or "average" Vietnam veterans to ascertain their success in post-combat adjustment or to understand the variables which influence readjustment over time. This I find extremely unfortunate because, without such incidence data, we will be unable to assess the size and scope of Vietnam veteran problems over the years and will therefore have great difficulty determining and documenting the quantity and quality of helping resources necessary to provide care that is needed. Such controlled quantitative longitudinal studies would also be helpful in lending perspective to the highly publicized media reports of uncontrolled studies of highly selected study samples. As I have reported elsewhere, I find it unfortunate that findings from samples of psychiatric patient veterans and antiwar veterans seeking help for adjustment difficulties have at times been generalized to apply to *all* Vietnam veterans—psychiatric patients and nonpatients, maladjusting and adjusting, antiwar and prowar in political feeling. I have found it equally distressing when the media have overgeneralized findings from my studies which, although controlled, are limited in their direct relevance to the military and civilian milieus of the early 1970s during the veterans' first 7 months of return. Again, controlled studies which would provide longitudinal perspective are desperately needed, even if now attempted retrospectively. To my mind, the only organizations with the mandate, access, and resources to undertake studies of sufficient magnitude would be federal-level governmental agencies.

It should always be kept in mind that any single description of the average or typical Vietnam veteran is a fiction. This is due primarily to the length of the war, the basic changes in the fabric and feelings of American society during those years, and the different single slice of the war experienced by each veteran. We must remember that the American combat participation in the war in Vietnam spanned over 10 years and three political administrations. Most veterans served only a 12- or 13-month tour of duty; each, therefore, experienced a different war during his particular year in his particular assignment at his particular level of combat or noncombat participation. Veterans returned to a changing United States, which had overpoweringly supported the war in its early years and made a massive shift in the late 1960s and early 1970s until, by the war's end in 1973, one was hard-pressed to find supporters of the war or even those willing to admit that they had supported it in the past (public citizen denial and projection of responsibility for the war in Vietnam has been a distressing phenomenon, leading to significant scapegoating of veterans and the military).

In short, to characterize all Vietnam veterans in aggregate without an enormously large sample containing subsamples of many types of veterans is scientifically unsound.

Lessons for the Military

What lessons can the military take from studies of Vietnam veteran readjustment that will be valuable in the future? First, it can acknowledge, belatedly, that the transition from the combat zone to the United States is a significantly stressful period and that manpower and monetary resources should be allocated for the planning and implementation of programs to facilitate reentry of combat veterans of future altercations. I hope that this will be a task of sufficient priority so that military behavioral scientists, joined by consultants from the civilian perspective, will begin at an early stage to plan interventions for reentering veterans, integrating the knowledge from Vietnam and earlier wars with the relevant specific issues of the combat zone and home front of the future. The Vietnam reentry studies should alert today's military to focus more attention on the reentry issues of soldiers making the transition from noncombat foreign assignments to the United States in which many of the same military, family, and at times social and emotional issues are relevant and stressful.

The second major lesson is that we have to learn to better prepare soldiers to cope with the mundane bureaucratic environment of garrison Army life. We prepare them well for the extreme environment of combat, but there is insufficient acknowledgment that most military roles are not martial in nature but center on maintaining a readiness for combat-related tasks which rarely occur. We must figure out how to make this prominent aspect of military garrison life less frustrating and how to make the garrison a milieu in which soldiers can find meaningful opportunities for both personal and military growth and development. This will mean finding meaningful (often not inherently "military") jobs in the peacetime volunteer Army for more men. We must learn to teach men how to maintain their self-esteem as well as to perform their tasks satisfactorily for themselves and for the service by helping them understand how to navigate our large military bureaucratic systems successfully.

We, as a country, did not fulfill our obligation to facilitate the reentry of the almost three million veterans who fulfilled their obligation to our country in the Vietnam combat zone and then returned to the stateside military to complete their active duty tours. I hope that some of the studies of the reentry transition from Vietnam will help us be better prepared and more willing to make helpful interventions into the expectably stressful military transitions of the future.

ADVOCACY AND CORRUPTION IN THE HEALING PROFESSIONS

Robert Jay Lifton

In looking at the professions, one does well, I think, to hold to the old religious distinction between the ministerial and the prophetic. One should not assume, as many do, a simple polarity in which the sciences are inherently radical or revolutionary and the healing professions intrinsically conservative. The professions must minister to people, take care of them, and that is a relatively conservative process. But there are prophets who emerge from the healing ministrations of the professions–Freud is a notable example–with radical critiques and revolutionary messages. Moreover, even "pure scientists" (in biology or physics, for example) spend most of their time ministering to the existing paradigm and strongly resist the breakthrough that is inevitably charted by the prophets among them. There are ministerial and prophetic elements in both the healing professions and the sciences.

But one must also distinguish between the professions, which have profound value in their capacity for continuity and renewal, and professional*ism,* the ideology of professional omniscience, which in our era inevitably leads to technicism and the model of the machine. The necessity for such a distinction becomes painfully clear if one looks at the situation that prevailed for psychiatrists in Vietnam. I want to take that situation as a starting point for a broader discussion of these dilemmas and their moral and conceptual ramifications. For the fact is that, in such extreme situations, the professional may be no better able than his soldier-patient to sort out the nuances of care and professional commitment, on the one hand, and moral (or immoral) action, on the other.

Central to my view of the present predicament of the professions is the psychology and the spirit of the survivor. The concept of the survivor derives in my work from the study I did in Hiroshima (Lifton, 1968), and has been fundamental to my subsequent thought. I define a survivor as one who has touched, witnessed, encountered, or been immersed in death in a literal or symbolic way and has himself remained alive. Let us assume that we as professionals share the national "death immersion" of not only Vietnam but also the related Watergate-impeachment process and that our ways of surviving them can have significance for us in our work. We may then discover that we are not entirely removed from the constellation of psychological patterns that I found Hiroshima survivors to share to a rather striking degree with survivors of Nazi death camps, the plagues of the Middle Ages, natural disasters, and what Kurt Vonnegut calls "plain old death."

I

The concept of the survivor includes five patterns. The first is the survivor's indelible death image and death anxiety. This "death imprint" often has to do with a loss of a sense of invulnerability. The second pattern, that of death guilt, is revealed in the survivor's classic question, "Why did I stay alive when he or she or they died?" The question itself has to do with a sense of organic social balance: "If I had died, he or she would have lived." That image of exchange of one life for another is perhaps the survivor's greatest psychological burden. A third pattern is that of desensitization or what I call psychic numbing, the breakdown of symbolic connectedness with one's environment. Numbing is a necessary protective mechanism in holocaust, but it can become self-perpetrating and express itself in sustained depression, despair, and apathy or withdrawal. A fourth survivor pattern has to do with the "death taint," as experienced by others toward survivors and by survivors themselves, resulting in discrimination against them and mutual suspicion and distrust. Central to this pattern is the survivor's "suspicion of counterfeit nurturance," his combination of feeling in need of help and resenting help offered as a reminder of weakness. (This kind of suspicion occurs not only in holocaust but in any situation in which "help" is offered by the privileged to the downtrodden or oppressed, as in white-black relations; and one can readily find models for this pained interaction in parent-child relationships.)

The fifth pattern is fundamental to all survivor psychology and encompasses the other four. It is the struggle to be able to find form and significance in one's remaining life experience. This formulative struggle is equally visible in more symbolic experiences of holocaust, such as surviving ways of life that one perceives to be "dying." In that sense, rapid social change makes survivors of us all.

In examining our healing professions in relation to these survivals, especially that of the psychiatric death encounter in Vietnam, we should keep in mind two general survivor alternatives. One can retreat from the issues raised by the death immersion and thereby remain bound to it in a condition of stasis (or numbing). Or one can confront the death immersion and derive insight, illumination, and change from the overall survivor experience (Lifton, 1973a). The latter response to some kind of experience of survival has probably been the source of most great religious and political movements, and of many breakthroughs in professional life as well.

A related issue is that of advocacy, which in our profession applies both to investigation and therapy, and is crucial to issues of professional renewal. I came to my work with Vietnam veterans from two directions, from prior antiwar advocacy and from professional concern with holocaust deriving from my research in Hiroshima. In the work with veterans I sought to combine detachment sufficient to enable me to make psychological evaluations (which I had to do at every step) with involvement that expressed my own commitments and moral passions. I believe that we always function within this dialectic between ethical involvement and intellectual rigor, and that bringing our advocacy "out front" and articulating it make us more, rather than less, scientific. Indeed, our scientific accuracy is likely to suffer when we hide our ethical beliefs behind the claim of neutrality and see ourselves as nothing but "neutral screens." The Vietnam War constitutes an extreme situation in which the need for an ethical response is very clear. But we have a tradition of great importance in depth psychology, much evident in Freud, of studying extremes in order to illuminate the (more obscure) ordinary.

II

I want to focus now on my experience over the last 3 years with "rap groups" of antiwar veterans and then to generalize from that experience about professional issues around advocacy and corruption.

The veterans' rap groups came into being because the veterans sensed that they had more psychological work to do in connection with the war (Lifton, 1973a). It is important to emphasize that the veterans themselves initiated the groups. The men knew that they were "hurting" but did not want to seek help from the Veterans Administration, which they associated with the military, the target of much of their rage. And though they knew they were in psychological pain, they did not consider themselves patients. They wanted to understand what they had been through, begin to heal themselves, and at the same time make known to the American public the human costs of the war. These two aspects of the veterans' aspirations in forming the groups—healing themselves while finding a mode of political expression—paralleled the professional dialectic of rigor and advocacy mentioned earlier. Without using those words, the veterans had that combination very much in mind when they asked me to work with them.

I called in other professionals (Chaim Shatan did much of the organizing) in the New York-New Haven area to assist us in the rap groups. I also participated with the veterans in the Winter Soldier Investigation of 1971, the first large-scale public hearing at which American GIs described their own involvement in war crimes. From the beginning the therapeutic and political aspects of our work developed simultaneously. It seemed natural for us to initiate the rap groups on the veterans' own turf, so to speak—in the office of the Vietnam Veterans Against the War—and to move to the neutral ground of a theological seminary when problems of space and political infighting developed at the VVAW office. The men wanted to meet where they were comfortable and where they could set the tone. With many stops and starts and much fluidity in general, a sizable number of these rap groups have formed, in New York City and throughout the country (Lifton, 1973a). The one that I have been part of has been meeting continually since the early 1970s.

We made plans for weekly 2-hour meetings, but the sessions were so intense, with such active involvement on the part of everybody, that they would generally run for 3 or 4 hours. I also interviewed many of the men individually in connection with the research I initiated then and subsequently published. From the beginning we avoided a medical model: we called ourselves "professionals" rather than "therapists" (the veterans often referred to us simply as "shrinks"), and we spoke of rap groups rather than group therapy. We were all on a first-name basis, and there was a fluidity in the boundaries between professionals and veterans. But the boundaries remained important nonetheless; distinctions remained important to both groups; and in the end the healing role of professionals was enhanced by the extent to which veterans could become healers to one another and, in some degree, to professionals also. Equally important, there was an assumption, at first unspoken and later articulated, that everybody's life was at issue; professionals had no special podium from which to avoid self-examination. We too could be challenged, questioned about anything—all of which seemed natural enough to the veterans but was a bit more problematic for the professionals. As people used to interpreting others' motivations, we first felt it a bit jarring to be confronted with hard questions about our own lives and with challenges about the way we lived. Not only was our willingness to share this kind of involvement crucial to the progress of the group, but in the end many of us among the professionals came to value and enjoy this kind of dialogue.

As in certain parallel experiments taking place, not only in psychological work but throughout American culture, we had a clearer idea of what we opposed (hierarchical distancing, medical mystification, psychological reductionism that undermines political and ethical ideas) than of what we favored as specific guidelines. But before long I came to recognize three principles that seemed important. The first was that of *affinity*, the coming together of people who share a particular (in this case overwhelming) historical or personal experience, along with a basic perspective on that experience, in order to make some sense of it (the professionals entered into this "affinity," at least to a certain extent, by dint of their political-ethical sympathies and inclination to act and experiment on behalf of them). The second principle was that of *presence*, a kind of being-there or full engagement and openness to mutual impact—no one ever being simply a therapist against whom things are rebounding. The third was that of *self-generation*, the need on the part of those seeking help, change, or insight of any kind to initiate their own process and conduct it largely on their own terms so that, even when calling in others with expert knowledge, they retain major responsibility for the shape and direction of the enterprise. Affinity, presence, and self-generation seem to be necessary ingredients for making a transition between old and new images and values, particularly when these relate to ultimate concerns, to shifting modes of historical continuity or what I have elsewhere called symbolic immortality (Lifton, 1973b).

I do not want to give the impression that everything went smoothly. There were a number of tensions in the group, one of them having to do with its degree of openness and fluidity. Openness was an organizing principle: in the fashion of "street-corner psychiatry," any Vietnam veteran was welcome to join a group at any time. Fluidity was dictated by the life styles of many of the veterans, who traveled extensively around the country and did not hold regular jobs. Professionals too were unable to attend every session. We established a policy of assigning three professionals to a group, with arrangements that at least one come to each meeting —but professionals became so involved in the process that there were usually two or all three present at a given group meeting.

When a veteran would appear at the group for the first time, obviously ill at ease, he would be welcomed by the others with a phrase, "You're our brother." Still, everyone came to recognize that such a policy could interfere with the probing of deep personal difficulties. A similar issue developed around the question of how accessible we would be to the media. Veterans wanted to make known the human costs of the war as part of their antiwar commitment; but after permitting a sympathetic journalist to sit in on a few of the sessions, they came to recognize that group process could be interfered with by the presence of even a sensitive outsider. We eventually arrived at the policy that only veterans and professionals could be present during group sessions, but that the group could on occasion meet with interested media people after a session was over. That solution served to protect the integrity of the group while conveying to journalists a rather vivid sense of both the impact of the war and the nature of the rap group experience.

There was also a tension among the professionals between two views of what we were doing. In the beginning a majority of the professionals felt that the essential model for our group sessions was group therapy. These professionals argued that the men were "hurting" and needed help and that, if we as therapists offered anything less than group therapy, we were cheating the men of what they most needed. I held to a second model which was at first a minority view. This position, while acknowledging the important therapeutic element of what we were doing, emphasized the experimental nature of our work in creating a new institution—a sustained dialogue between professionals and veterans, based on a common stance of opposition to the war, in which both drew upon their special knowledge, experience, and needs. This model did not abolish role definitions: veterans were essentially there to be helped and professionals to help, but it placed more stress on mutuality and shared commitment.

We never totally resolved the tension between these two models in the sense that all of us came to share a single position. The veterans tended to favor the second model but did not want to be shortchanged in terms of the help they wanted and needed. There was a continuing dialectic between these two ways of seeing what we were doing. But those who held to the second model—which related to other experiments taking place in American society with which the veterans identified—tended to stay longer with the project. Those professionals who conceived the effort in a more narrowly defined therapeutic way, and who I suspect were less politically and ethically committed to an antiwar position, tended to leave.

Of course, there were differences in professional style even within these two models. Some professionals were particularly skilled at uncovering the childhood origins of conflicts. I was seen as an authority on issues of death and life continuity and on social-historical dimensions around death and survival. As a personal style, my impulse was to be something of a mediator, and the group soon came to see me in that way. The group recognized and accepted differing personal styles, and it was interesting to us as professionals to observe reflections of ourselves in the responses to us not only of veterans but also of other professionals sharing the experience of a particular group. The veterans began with almost no knowledge of group process, but they learned quickly. From the beginning the focus of the group was on the overwhelming experience of the war and on residual guilt and rage. In the process of examining these issues, the men looked increasingly at their ongoing life struggles, especially their relationships with women, feelings about masculinity, and conflicts around work. There was a back-and-forth rhythm in the group between immediate life situations and war-related issues, with these gradually blending in deepening self-examination that was generally associated in turn with social and political forces in the society.

For all of us in the group there was a sense that the combination of ultimate questions (around death and survival) and experimental arrangements required that we call upon new aspects of ourselves and become something more than we had been before. Central to this process was the changing relationship between veterans and professionals. At moments the veterans could become critical of the way professionals were functioning. There were, of course, may conflicts, but there was also an essential feeling of moving toward authentic insight. By taking seriously the issues as they were raised, we maintained a double level of individual-psychological interpretation and shared actuality. Taking that actuality seriously contributed to the sense of everyone's "presence."

III

The rap groups represented a struggle on the part of both veterans and psychological professionals to give form to what was in many ways a common survival, a survival for the veterans of a terrible death immersion and for the professionals of their own dislocations in relationship to the war and the society. During our most honest moments, we professionals have admitted that the experience has been as important for our souls as for theirs. For the rap groups have been one small expression (throughout the country they and related programs have probably involved, at most, a few thousand people) of a much larger cultural struggle toward creating what I have termed animating institutions (Lifton, 1973a, 1973c). Whether these emerge from existing institutions significantly modified or as "alternative institutions," they can serve the important function of providing new ways of being a professional as well as of relating to professionals. While such institutions clearly have radical possibilities, they can also serve a genuinely conservative function in enabling those involved to find a means of continuing to relate, however critically, to the existing society and its other institutions, as opposed to retreating in embittered alienation, destructiveness, or self-destructiveness. In this and other ways, the rap group experience seemed to me a mirror on psychohistorical struggles of considerable importance throughout the society.

A compelling example was the rap group's continuous preoccupation with struggles around maleness. I described these struggles in a chapter of my book entitled, somewhat whimsically but not without seriousness, "From John Wayne to Country Joe and the Fish." The men were very intent upon examining what they came to call "the John Wayne thing" in themselves—a process actively encouraged if not required by girl friends and wives often active in the women's movement. The essence of the issue for the veterans was their deepening realization that various expressions of supermaleness encouraged in American culture were inseparable from their own relationship to war making. They probed unsparingly the sources and fears beneath their male bravado in enthusiastically (in many cases) joining up and even seeking out the war. For the insight that graudally imposed itself on them was that only by extricating themselves from elements of "the John Wayne thing"—notably its easy violence on behalf of unquestioned group loyalty and its male mystique of unlimited physical prowess always available for demonstration—could they, in a genuine psychological sense, extricate themselves from the war. Two significant psychological alternatives were available to them from the youth culture of the 1960s: the image of a male's being no less genuinely so for manifesting tenderness, softness, aesthetic sensitivity, and awareness of feelings, and the overall social critique of war, war making, and the warrior ethos. It was particularly the latter that Country Joe MacDonald and his rock group (Country Joe and the Fish) gave ecstatic expression to in their celebrated song, "I Feel Like I'm Fixin' To Die Rag." A frenzied and bitter evocation of the absurdity of dying in Vietnam, the song propels one to the far reaches of the grotesque: "And it's 1, 2, 3, what are we fighting for?/Don't ask me I don't give a damn. . . . Well there ain't no time to wonder why/Whoopee we're all gonna die." And a little later in the song: "Well come on mothers throughout the land/Pack your boys off to Vietnam/Be the first one on your block/To have your boy come home in a box." Ironically and significantly, the "Fixin' To Die Rag" was probably the song most frequently played by men serving in Vietnam, and it is very likely that this expression of the utter absurdity and grotesqueness of dying in Vietnam will become *the* song of the Vietnam War.

The personal transformation that the veterans experience (barely suggested here) can thus be seen to have both introspective and extrospective elements. The men constantly look inward, but they also look outward at their society in relationship both to having been drawn into the war and to what they perceived as a dubious welcome upon their return. This extrospective aspect of personal change is always important—not only in experimental institutions like the rap groups but also in ordinary psychotherapy and ordinary living—but is often denied or ignored because of the implicit assumption that psychological experience, being internal, is totally self-contained. It was precisely this dual vision that enabled many veterans to develop what I came to see as an animating relationship to their sense of guilt (Lifton, 1973a). In contrast to static (neurotic) forms of guilt and immobilizing self-condemnation, animating guilt can provide energy toward change via the capacity to examine the roots of that guilt in both social and individual terms. I believe that these distinctions around guilt, when pursued further, have significance both for depth psychological theory and for the ethnical questions at issue in this discussion (Lifton, 1972).

IV

Guilt and rage were fundamental emotions that we explored constantly in the groups. But the men had a special kind of anger best described as ironic rage toward two types of professionals with whom they came into contact in Vietnam, chaplains and "shrinks." They talked about chaplains with great anger and resentment as having blessed the troops, their mission, their guns, and their killing: "Whatever we were doing—murder, atrocities—God was always on our side." Catholic veterans spoke of having confessed to meaningless transgressions ("Sure, I'm smoking dope again. I guess I blew my state of grace again") while never being held accountable for the ultimate one ("But I didn't say anything about killing"). It was as if the chaplains were saying to them, "Stay within our moral clichés as a way of draining off excess guilt, and then feel free to plunge into the business at hand."

The men also pointed to the chaplain's even more direct role of promoting false witness. One man spoke especially bitterly of "chaplains' bullshit." He illustrated what he meant by recalling the death of a close buddy—probably the most overwhelming experience one can have in combat—followed by a combined funeral ceremony-pep talk at which the chaplain urged the men to "kill more of them." It is not generally recognized that the My Lai massacre occurred immediately after the grotesque death—caused by an exploding booby trap—of a fatherly, much revered noncommissioned officer that had been witnessed by many of the men. His funeral ceremony was conducted jointly by a chaplain and the commanding officer, the former blending spiritual legitimacy to the latter's mixture of eulogy and exhortation to "kill everything in the village." A eulogy in any funeral service asks those in attendance to carry forward the work of the dead person. In war, that "work" characteristically consists of getting back at the enemy, thereby providing men with a means of resolving survivor guilt and with a "survivor mission" involving a sense not only of revenge but also of carrying forth the task that the fallen comrade could not see to completion. In Vietnam, the combination of a hostile environment and the absence of an identifiable enemy led to the frequent manipulation of grief to generate a form of false witness, a survivor mission of atrocity (Lifton, 1973a).

The men spoke with the same bitterness about "shrinks" whom they had encountered in Vietnam. They described situations in which they or others experienced an overwhelming combination of psychological conflict and moral revulsion, difficult to distinguish in Vietnam. Whether one then got to see a chaplain, a psychiatrist, or an enlisted-man assistant of either had to do with where one was at the time, who was available, and the attitudes of the soldier and the authorities in his unit toward religion and psychiatry. But should he succeed in seeing a psychiatrist, he was likely to be "helped" to remain at duty and (in many cases) to carry on with the daily commission of war crimes, which was what the ordinary GI was too often doing in Vietnam. Psychiatry for these men served to erode whatever capacity they retained for moral revulsion and animating guilt. They talked in the rap groups about ways in which psychiatry became inseparable from military authority.

But in their resentment toward chaplains and psychiatrists the men were saying something more. It was one thing to be ordered by command into a situation that they came to perceive as both absurd and evil, but it was quite another to have that process rationalized and justified by ultimate authorities of the spirit and mind—that is, by chaplains and psychiatrists. The chaplains and psychiatrists formed an unholy alliance not only with the military command but also with the more corruptible elements in the soldier's psyche, corruptible elements available to all of us, whether soldier or chaplain or psychiatrist.

We can thus speak of the existence of a "counterfeit universe" in which pervasive, spiritually reinforced inner corruption becomes the price of survival. In this sense, the chaplains and psychiatrists were just as entrapped as the GIs. For we may assume that most of them were reasonably conscientious and decent professionals—much like the writer and reader of this article—caught up in an institutional commitment in this particular war.

When in our group the men spoke harshly of military psychiatrists, we professionals of course asked ourselves whether they were talking about us. In some degree they undoubtedly were. They were raising the question of whether *any* encounter with a psychiatrist, even in a context which they themselves created and into which we were called, could be any more authentic than the counterfeit moral universe to which psychiatrists had lent themselves in Vietnam.

V

I want to move now to some reflections about psychiatry in more ordinary situations. The rap group experience raised questions about the extent to which everyday work in our profession, and in the professions in general, tends to wash away rather than pursue fundamental struggles around integrity—the extent to which the special armor of professionals blocks free exchange between them and the people they intend to serve.

In the rap group experience I found the issue of investigative advocacy more pressing and powerful than in other research I have done.[1] This was partly because veterans and professionals alike were more or less in the middle of the problem—the war continued and we all had painful emotions about what it was doing and what we were doing or not doing to combat it.

But I also came to realize that, apart from the war, the work had important bearing upon a sense of long-standing crisis affecting members of all the professions—a crisis that the war in Vietnam both accentuated and illuminated but by no means created. We professionals, in other words, came to the rap groups with our own need for a transformation in many ways parallel to if more muted than that which we sought to enhance in veterans. We too, sometimes with less awareness than they, were in the midst of struggles around living and working that had to do with intactness and wholeness, with what we have been calling integrity.

One source of perspective on that struggle, I found, was a return to the root ideas of profession, the idea of what it means to profess. Indeed, an examination of the evolution of these two words could provide something close to cultural history of the West. The Latin prefix *pro* means forward, toward the front, forth, out, or into a public position. *Fess* derives from the Latin *fateri* or *fass,* meaning to confess, own, acknowledge. To profess (or be professed), then, originally meant a personal form of out-front public acknowledgment. And that which was acknowledged or "confessed" always (until the 16th century) had to do with religion, with taking the vows of a religious order or declaring one's religious faith. But as society became secularized, the word came to mean "to make claim to have knowledge of an art or science" or "to declare oneself expert or proficient in" an enterprise of any kind. The noun form, *profession,* came to suggest not only the act of professing but also the ordering, collectivization, and transmission of the whole process. The sequence was from *profession* or religious conviction (from the 12th century) to a particular order or *professed persons,* such as monks or nuns (14th century) to "the occupation which one possesses to be skilled in and follow," especially "the three learned professions of divinity, law, and medicine," along with the "military profession." So quickly did the connotations of specialization and application take hold that as early as 1605 Francis Bacon could complain: "Amongst so many great foundations of colleges in Europe, I find strange that they are all dedicated to professions, and none left free to Art and Sciences at large" (Lifton, 1973a).

Thus the poles of meaning around the image of profession shifted from the proclamation of personal dedication to transcendent principles to membership in and mastery of a specialized form of socially applicable knowledge and skill. In either case the profession is immortalizing–the one through the religious mode, the other through works and social-intellectual tradition. And the principles of public proclamation and personal discipline carry over from one meaning to the other–the former taking the shape of examination and licensing; the latter, of study, training, and dedication. Overall, the change was from advocacy based on faith to technique devoid of advocacy.

To be sure, contemporary professions do contain general forms of advocacy: in law, of a body of suprapersonal rules applicable to everyone; in medicine, of healing; and in psychiatry, of humane principles of psychological well-being and growth. But immediate issues of value-centered advocacy and choice (involving groups and causes served and the consequences thereof) are mostly ignored. I am not advocating a return to pure faith as a replacement for the contemporary idea of what profession means. But I am suggesting that the notion of profession should include these issues of advocacy and ethical commitment. The psychiatrist in Vietnam, for example, whatever his intentions, found himself in collusion with the military in conveying to individual GIs an overall organizational message: "Do your indiscriminate killing with confidence that you will receive expert medical-psychological help if needed."

Three well-known principles of military psychiatry developed during recent wars are *immediacy* (a soldier is treated as soon as possible), *proximity* (close to the combat area), and *expectancy* (everyone under treatment is from the beginning made to expect that he will return to duty with his unit). There is a certain logic to these principles. Their use very often does eliminate or minimize the secondary gains from illness and the chronic symptomatology that would otherwise ensue when men are sent to the rear to undergo prolonged psychiatric hospitalization, as well as feelings of failure and unmanliness that become associated with eventual medical discharge from the military. One psychiatric report from Vietnam describes the use of these principles and the assumption that those requiring treatment "had run into some difficulty in interpersonal relations in their units that caused them to be extruded from these groups," so that "The therapeutic endeavor . . . was to facilitate the men's integration into their own groups (units) through integration into the group of ward patients" (Bloch, 1969).

The approach seems convincing until one evaluates some of the conditions under which atrocities occurred or were avoided. I spent 10 years interviewing a man who had been at My Lai and had not fired or even pretended he was firing. (Among the handful who did not fire, most held their guns in position as if firing in order to avoid the resentment of the majority actively participating in the atrocity.) Part of what sustained this man and gave him the strength to risk ostracism was his very distance from the group. Always a loner, he had been raised near the ocean and as a child engaged mainly in such solitary activities as boating and fishing. Hence, though an excellent soldier, he was less susceptible than others to group influence and in fact remained sufficiently apart from other men in his company to be considered "maladapted" to that immediate group situation (Lifton, 1973a). One must distinguish between group integration and personal integrity—the latter including moral and psychological elements that connect one to social and historical context beyond the immediate. Group integration can readily undermine integrity—in Vietnam, for both the soldier and the psychiatrist who must grapple with his own struggles to adapt to a military institution with its goals of maximum combat strength and to a combat situation of absurdity and evil. No wonder that, in Vietnam, he found little ethical space in which to move. The clear implication here is that the psychiatrist, no less than the combat soldier, is confronted with the important question of the group that he is to serve and, above all, the nature and consequences of its immeidate and long-range mission. To deal with that, he must overcome the technicist assumption we fall into all too easily, namely, "Because I am a healer, anything I do, anywhere, is good." It may not be.

VI

I wonder how many colleagues shared my sense of chilling illumination in picking up the October 1971 issue of the *American Journal of Psychiatry* and finding in it two articles by psychiatrists writing about Vietnam: one entitled "Organizational Consultation in a Combat Unit" (Bey and Smith, 1971) and the other "Some Remarks on Slaughter" (Gault, 1971). The first, by Douglas R. Bey, Jr., and Walter E. Smith, lives up to its title in providing a military-managerial view of the psychiatrist's task. The authors invoke a scholarly and "responsible" tone as they describe the three principles of combat psychiatry and trace their historical development. They then elaborate their own "workable method of organizational consultation developed and employed in a combat division in Vietnam." The method combines these principles of military psychiatry with "an organizational case study method" recently elaborated for industry at the Menninger Foundation and, according to the authors, has bearing on possible developments in community psychiatry. Their professional voice sounds tempered, practical, and modest as they tell of their team approach (with trained corpsmen) and of interviews with commanding officers, chaplains, and influential noncoms and as they acknowledge that commanders "were far better prepared to work out solutions to their problems than we, since their area of expertise was in administration and fighting whereas ours was in the area of helping them to see where their feelings might be interfering with their use of these skills."

It was enough for psychiatric consultants to serve as an "observing ego" to the particular military unit. To back up that position they quote, appropriately enough, from an article by General W.C. Westmoreland recommending that the psychiatrist assume "a personnel management consultation type role." The title of that article by General Westmoreland, "Mental Health–An Aspect of Command," makes quite clear just whom psychiatry in the military is expected to serve.

The authors' combination of easy optimism and concern for everyone's feelings, and for the group as a whole, makes one almost forget the kinds of activities in which the members of that group were engaged. Reading that lead article in the official journal of the national organization of American psychiatrists gave me a disturbing sense of how far this kind of managerial technicism could take a profession, and its reasonably decent individual practitioners, into ethical corruption. What is most significant about the article is that the authors never mention the slightest conflict–in themselves and their psychiatric team any more than in the officers and men they dealt with–between group integration and personal integrity. Either they were too numbed to be aware of such conflict or (more likely) they did not consider it worthy of mention in a scientific paper.

William Barry Gault's article, "Some Remarks on Slaughter," was a particularly welcome antidote, even if a bit more hidden in the inside pages. As his title makes clear, Gault's tone is informed by an appropriate sense of outrage. Significantly, his vantage point was not Vietnam but Fort Knox, Kentucky, where he examined large numbers of men returning from combat. He was thus free of the requirement of integration with a combat unit, and we sense immediately a critical detachment from the atrocity-producing situation.

Gault introduces the idea of "the psychology of slaughter," combining the dictionary definition of that word ("the extensive, violent, bloody or wanton destruction of life; carnage") with a psychological emphasis upon the victim's defenselessness ("whether . . . a disarmed prisoner or an unarmed civilian"). He can "thus . . . distinguish slaughter from the mutual homicide of the actual combatants in military battle." He sets himself the interpretive task of explaining how "relatively normal men overcame and eventually neutralized their natural repugnance toward slaughter." He is rigorously professional as he ticks off six psychological themes or principles contributing to slaughter, and yet his ethical outrage is present in every word. His themes are: "The enemy is everywhere" or "the universalization of the enemy"; "the enemy is not human" or "the 'cartoonization' of the victim"; the "dilution" or "vertical dilution" of responsibility; the "pressure to act"; "the natural dominance of the psychopath"; and "sheer firepower . . . [such that] terrified and furious teenagers by the tens of thousands have only to twitch their index finger, and what was a quiet village is suddenly a slaughterhouse."

Gault sensitively documents each of these themes in ways very consistent with experiences conveyed to me in rap groups and individual interviews. He ends his article with illustrative stories: prisoners who refused to give information were thrown out of helicopters as examples to others; a new combat commander who refused to shoot a twelve-year-old "dink" accidentally encountered by a company while setting up an ambush drew the hostility of his own men, who saw the whole company jeopardized by the survival of someone who might, even as a prisoner, convey information about the ambush. Gault admits he does not know "why similar experiences provoke so much more guilt in one man than in another"; and, still professionally cautious (perhaps overly so), he remains "unwilling to attempt to draw any large lessons" from his observations. At the end he insists only that "in Vietnam a number of fairly ordinary young men have been psychologically ready to engage in slaughter and that moreover this readiness is by no means incomprehensible."

One senses that these stories made a profound impact upon Gault, that he became a survivor of Vietnam by proxy, and that the article was his way of giving form to that survival as well as resolving his own integration-integrity conflict as a morally sensitive psychiatrist in the military at the time of the Vietnam War.[2] He was able to call forth his revulsion toward the slaughter (and, by implication, his advocacy of life-enhancing alternatives) as a stimulus to understanding and to bring to bear on the Vietnam War a valuable combination of professional insight and ethical awareness.

Nor do we have follow-up studies on psychiatrists and their spiritual-psychological state after service in Vietnam. I have talked to a number of them, and my impression is that they find it no easier to come to terms with their immersion in the counterfeit universe than does the average GI. They too feel themselves deeply compromised. They seem to require a year or more to begin to confront the inner contradictions that they have experienced. They too are survivors of

Vietnam, and of a very special kind. I know of one or two who have embarked upon valuable survivor missions, parallel to and partly in affiliation with that of VVAW as an organization. But what is yet to emerge, though I hope it will before too long, is a detailed personal account by a psychiatrist of his struggles with group integration and individual integrity, and with the vast ramifications of the counterfeit that this paper only begins to suggest.

Put simple, American culture has so technicized the idea of psychiatric illness and cure that the psychiatrist or psychoanalyst is thrust into a stance of scientifically based spiritual omniscience—a stance he is likely to find much too seductive to refuse entirely. Anointed with both omniscience and objectivity and working within a market economy, his allegedly neutral talents become available to the highest bidder. In a militarized society they are equally available to the war makers.

VII

An alternative perspective, in my judgment, must be not only psychohistorical, but also psychoformative. By the latter I mean a stress upon the process of inwardly re-creating all that is perceived or encountered (Lifton, 1973b, n.d.). My stress is upon what can be called a formative-symbolic *process,* upon symbolization rather than any particular symbol (in the sense of one thing's standing for another). The approach connects with much in 20th century thought and seeks to overcome the 19th century emphasis upon mechanism, with its stress upon breakdown of elements into component parts—an emphasis inherited, at least in large part, by psychoanalysis, as the word itself suggests.

The antiwar passions of a particular Vietnam veteran, on the contrary, have to be understood as a combined expression of many different psychic images and forms: the Vietnam environment and the forces shaping it; past individual history; the post-Vietnam American experience, including VVAW and the rap groups and the historical forces shaping these; and the various emanations of guilt, rage, and altered self-process that could and did take shape. Moreover, professionals, like myself, who entered into the lives of these veterans—with our own personal and professional histories, personal struggles involving the war, and much else—became a part of the overall image-form constellation.

Psychiatrists have a great temptation to swim with an American tide that grants them considerable professional status but resists, at times quite fiercely, serious attempts to alter existing social and institutional arrangements. As depth psychologists and psychoanalysts we make a kind of devil's bargain that we can plunge as deeply as we like into intrapsychic conflicts while not touching too critically upon historical dimensions that question those institutional arrangements. We often accept this dichotomy quite readily with the rationale that, after all, we are not historians or sociologists. But the veterans' experience shows that one need extrospection as well as introspection to deal with psychological conflicts, particularly at a time of rapid social change. And I believe that a general psychological paradigm of "death and the continuity of life" (Lifton, n.d.) helps us to achieve this dual perspective and to recognize the interplay of psychological and moral elements in relationship to ultimate commitments and our own involvement in that interplay.

All this points toward the need for a transformation of the healing professions themselves. In my work with veterans, I restated a model of change that I had elaborated in earlier work, based on a sequence of confrontation, reordering, and renewal (Lifton, 1973a, 1961). The idea is worth stating at least as a model—not with any expectation of instant transformation, but with the recognition that, here and there, people are already pursuing it and will undoubtedly continue to do so in forms we have not yet imagined. Confrontation for the veterans meant confronting the idea of dying in Vietnam, often through the death of a buddy. For psychiatrists, it would mean confronting our own concerns about death, mortality, and immortality and our professional struggles with them. Reordering for veterans meant the working through of difficult emotions around guilt and rage; for psychiatrists this would mean seeking animating relationships to the same emotions in ourselves and recognizing and making use of our experience of despair. Renewal for veterans meant a new sense of self and world, including an enhanced playfulness; the professional parallels are there, and much can be said for the evolution of more playful modes of investigation and therapy.

I want to conclude with two quotations. The first is from Stanley Milgram, who performed controversial experiments on the willingness of people to cause pain and even endanger the lives of others when authoritatively requested to do so. Whatever one's views of the scientific and moral aspects of these "Eichmann experiments," one of Milgram's own conclusions is worth thinking about: "Men are doomed if they act only within the alternatives handed down to them" (Milgram, 1963, 1967, 1974).

And finally, Joseph Campbell, perhaps America's most distinguished student of mythology: "A god outgrown becomes immediately a life-destroying demon. The form has to be broken and the energies released" (Campbell, 1956).

Notes

1. In contrast, my Hiroshima work, in which I also experienced strong ethical involvement, was retrospective and in a sense prospective (there were immediate nuclear problems, of course, but we were not in the midst of nuclear holocaust); my study of Chinese thought reform dealt with matters having immediate importance but going on (in a cultural sense) far away; and my work with Japanese youth had much less to do with overwhelming threat and ethical crisis (1970a, 1970b, 1961).

2. This assumption that the article was an expression of Gault's own survivor formulation, which I made originally only on the basis of reading it (and, of course, on my experience, personal and professional, with the psychology of the survivor), was strongly confirmed by a brief talk he and I had when we met as members of a panel on Vietnam veterans at a psychoanalytic conference.

III.

REACTIONS TO THE ORGANIZATIONAL ENVIRONMENT

HEALTH AND ADJUSTMENT OF MEN AT SEA

E.K. Eric Gunderson

This paper considers the distributions of mental and physical illnesses in the U.S. naval population, some general factors that affect illness rates, and some specific aspects of the shipboard environment that are related to health and adjustment.

Navy ships offer distinct advantages for comparative studies of illness and other behavioral criteria. They are natural ecological units in which crew members experience the same food sources, water supplies, and general environmental conditions while at sea. Naval organizations tend to have well-defined occupational and social structures; crews tend to be homogeneous with respect to age and social background, comparable from one ship to another, and fairly stable in composition during overseas deployments. Medical services are readily available, and illness recording is standardized.

During the past several years Saul Sells, with the collaboration of the author, has been developing an approach to the comparative analysis of organizations in terms of a "social systems model" (Gunderson and Sells, 1975; Sells and Gunderson, 1972). The model provides a framework for analyzing relationships between the individual and his environment in an organizational context and also for evaluating the effectiveness of individuals, organizational units, and the organization as a whole in achieving organizational objectives.

The application of the social systems approach to analysis of naval organizations requires careful consideration of the sociocultural and political contexts in which the modern naval establishment operates. The Navy may be viewed as one of many technologically advanced programs in modern society. It is also a self-contained institution which has a unique history and traditions. The naval organization is

AUTHOR'S NOTE: *Report Number 75-19, supported by the Bureau of Medicine and Surgery, U.S. Department of the Navy, under Research Work Unit MF51.524.002-0003 and by the Office of Naval Research Contract Numbers ONR N00014-72-A-0179-0001 and RR042-08-01 NR170-743. Opinions expressed are those of the author and are not to be construed as necessarily reflecting the official view or endorsement of the U.S. Department of the Navy.*

faced with special personnel problems because of the mobility of naval units, which results in the disruption of normal living patterns, extended periods of separation from families and home communities, and confinement and restricted activities while at sea. Shipboard living may involve excessive noise, crowding, heat stress and poor ventilation, unpleasant odors, lack of privacy, lack of recreational facilities, boredom, long or irregular working hours with sleep deprivation, poor environmental design (for example, mazes of pipes, wiring, and ducts in living spaces), and arduous and incessant routine maintenance (cleaning, painting, and repair). The effects of these types of stresses upon health, work efficiency, job satisfaction, and career motivation have not been systematically studied. Also, the effects of changes in technology, ship design, and mission requirements, as well as basic life-support elements, such as food preparation, sanitation, recreational facilities, medical support, communications, and other factors, need to be evaluated in order to achieve an adequate understanding of naval organizations as social systems.

Physical illnesses and accidents, psychiatric disorders and behavioral problems, and low retention rates have long been recognized as major drains on human resources in the military services. A series of epidemiological and longitudinal studies concerned with manpower losses from illnesses, accidents, behavior problems, and premature attrition for psychiatric reasons have been conducted at the Naval Health Research Center, San Diego (Gunderson, 1971; Gunderson and Rahe, 1974; Rahe et al., 1972). These studies have provided basic information on health and personnel effectiveness of naval personnel and identified a number of correlates of health and adjustment.

Mental Health of Naval Personnel

Mental and emotional disorders represent a major public health problem in the armed forces as they do in society generally. During a peacetime period prior to the Vietnam War (1960-1962) approximately 9,000 new cases of diagnosed psychiatric disorders were admitted to naval medical facilities each year; for this period, psychiatric disorders were the leading cause of invaliding (medically discharging) personnel from the naval service, and nearly one million man-days were lost because of hospitalizations for psychiatric illness.

During the Vietnam War the psychiatric incidence rate for Navy enlisted men remained stable at 1,000 per 100,000 strength per year, but the rate for Marine Corps enlisted personnel more than doubled—from 1,000 in 1966 to about 2,100 per 100,000 per year in 1969. During 1968 and 1969, Marines were involved in intense and sustained fighting and suffered heavy battle casualties. The psychiatric incidence rate in Vietnam was considerably lower than that for World War II or Korea. This has been attributed to the clearly defined and relatively short tours of combat duty, the intermittent nature of combat operations, the excellent morale and motivation of the troops, and the efficiency of medical and psychiatric support in the field (Gunderson, 1971).

Psychiatric incidence rates varied widely in the Navy population by rank (288 per 100,000 per year for male officers versus 1,000 per 100,000 per year for male enlisted), sex (approximately 4,000 per 100,000 per year for female enlisted versus 1,000 per 100,000 for male enlisted), and age (approximately 2,400 per 100,000 for age 17-18 male enlisted versus approximately 750 per 100,000 for age 21-35 male enlisted).

Rates of psychiatric disorder in male Navy enlisted personnel were computed for 29 occupational specialties during fiscal year 1966-1967 (Gunderson, 1971). A summary of occupational differences using the device of comparing expected numbers of cases (based upon Navy-wide strengths and patient counts by pay grade) with actual numbers of cases during the two-year period is shown in Table 1.

Generally, men in technical jobs (radarman, fire control technician, electronics technician, and communication technician groups) had well below the expected number of psychiatric cases (100%). Also, aviation rating groups, with the exception of aviation boatswain's mate, tended to have fewer psychiatric admissions than the Navy as a whole.

The highest incidence rate was for hospital corpsmen; this group had more than double the expected number of cases. The next highest incidence rate was obtained for the boatswain's mate group. Boatswain's mates who had not achieved petty officer status or had lost petty officer status because of disciplinary actions had especially high rates of mental illness in all diagnostic categories, but particularly in the psychoses.

A comparison of psychiatric incidence rates for various types of

Table 1. EXPECTED AND ACTUAL INCIDENCE OF PSYCHIATRIC DISORDERS BY OCCUPATIONAL RATING GROUP*

	Number of Psychiatric Disorders		
Enlisted Ratings	**Expected[a]**	**Actual[b]**	**Percentage of Expected[c]**
Deck Group			
Boatswain's Mate	173	331	191.3
Radarman	150	101	67.3
Quartermaster	71	64	90.1
Ordnance Group			
Torpedoman's Mate	73	86	117.8
Gunner's Mate	131	108	82.4
Fire Control Technician	147	100	68.0
Electronics Group			
Electronics Technician	248	146	58.9
Administrative and Clerical			
Radioman	316	269	85.1
Communications Technician	159	73	45.9
Yeoman	230	192	83.5
Storekeeper	154	153	99.4
Commissaryman	121	185	152.9
Ship's Serviceman	65	67	103.1
Personnelman	85	111	130.6
Engine and Hull Group			
Machinist's Mate	349	345	98.8
Engineman	177	170	96.0
Boilerman	166	186	112.0
Electrician's Mate	237	205	86.5
Interior Communication Electrician	113	92	81.4
Shipfitter	99	81	81.8
Construction Group			
Equipment Operator	53	56	105.7
Aviation Group			
Aviation Machinist's Mate	277	160	57.8
Aviation Electronics Technician	150	102	68.0
Aviation Structural Mechanic	198	170	85.8
Aviation Boatswain's Mate	82	75	91.5
Aviation Electrician's Mate	110	69	62.7
Aviation Ordnanceman	81	61	75.3
Medical Group			
Hospital Corpsman	478	1,077	225.3
Steward Group			
Steward	290	105	36.2

*Reprinted from pp. 192-193 of C.D. Spielberger (ed.), *Current Topics in Clinical and Community Psychology,* Vol. 3 (New York: Academic Press, 1971).

a. The expected number of cases is based upon Navy-wide incidence rates for each pay grade during fiscal year 1966-1967.

b. All psychiatric cases for the designated rating group during fiscal year 1966-1967.

c. The percentage of expected is based upon the actual number of cases divided by the expected number of cases times 100. The percentage reflects the numbers of cases in particular rating groups relative to the numbers of cases in the total Navy for comparable pay grades.

Table 2. PSYCHIATRIC INCIDENCE RATES FOR SELECTED TYPES OF COMBAT SHIPS DURING THE VIETNAM WAR

Ship Type	Designator	Number of Ships	Incidence Rate[a]
Surface Warships:			
Aircraft Carrier, Attack and Support	CVA, CVS	22	1,323
Cruiser, Heavy and Guided Missile	CA, CG, CGN	7	801
Frigate, Guided Missile	LDG, DLGN	26	859
Destroyer	DD	130	1,115
Destroyer (Training)	DD	30	1,484
Destroyer, Guided Missile	DDG	29	971
Destroyer Escort	DE	14	676
Destroyer Escort (Training)	DE	19	1,057
Destroyer Escort, Guided Missile	DEG	6	702
Minesweeper	MSO, MSC	58	1,089
Minesweeper (Training)	MSO, MSC	23	1,702
Submarines:			
Diesel-electric, Auxiliary	SS, AGSS	61	577
Nuclear, Attack	SSN	47	458
Nuclear, Ballistic Missile	SSBN	39	344

a. The incidence rate is the number of new cases per 100,000 strength per year from July 1965 through December 1971.

combat ships during the Vietnam War is shown in Table 2. Aircraft carrier crews tended to have relatively high incidence rates while submarines, particularly the ballistic missile type, had very low rates. The destroyers, many of them old World War II types, had slightly higher than normal incidence rates (approximately 1,000), while destroyer escorts, many of them new ships of modern design, tended to have lower rates. Ships designated for training functions had high psychiatric incidence rates compared to operational ships of the same type.

Actually, the highest shipboard psychiatric incidence rate during the Vietnam War was not found among combat ship crews but among hospital ship crews and medical staff personnel (1,936 per 100,000 per year). These hospital ships received a heavy flow of casualties directly from the combat zone over extended periods of time (Strange and Arthur, 1967). As seen in Table 1, hospital corpsmen generally had the highest rate of psychiatric admissions of any Navy occupational group.

Shipboard Morbidity Studies

During 1967-1969, large-scale prospective studies were conducted aboard 3 cruisers, 2 aircraft carriers, and 1 battleship during 6- to 8-month overseas deployments (Doll et al., 1969; Gunderson et al., 1970; Rahe, Gunderson, and Arthur, 1970; Rahe, Mahan, Arthur, and Gunderson, 1970; Rubin, Gunderson, and Arthur, 1969, 1971a, 1971b; Rubin, Gunderson, and Doll, 1969). In these studies, illness rates (dispensary visits) were shown to vary consistently with age, pay grade, military experience, race, marital status, education, General Classification Test scores, occupational specialty, job satisfaction, self-perceptions of health, recent stressful life changes, and ship's operational activities. One salient finding in these studies was that younger, less experienced men had much higher illness rates than older, experienced sailors. Also, men working in physically demanding and hazardous environments—that is, blue-collar jobs, such as engineering, deck, and ordnance—had relatively high illness and injury rates. Similar results have been reported in civilian industry.

In these earlier studies, it was striking that overall illness rates varied considerably among the ships studied, even ships of the same type. These differences in illness rates could not be explained by differences in crew composition, operating schedules, or illness-reporting procedures. When the effect of a specific ship on the prediction-of-illness rate was evaluated using multiple regression procedures, an individual ship was found to contribute substantially to the prediction-of-individual-illness rates, suggesting that organizational and social factors aboard such ships might be important (McDonald et al., 1973). Thus, it was hypothesized that a combination of environmental variables (living and working conditions) and organizational and social context variables, as well as interactions among these factors, would account for additional variance in morbidity rates aboard Navy ships. Further research was needed to identify such additional sources of variance in illness rates among ships and other types of organizations and to evaluate their relative importance. In 1972 new studies, supported by the Bureau of Medicine and Surgery and the Office of Naval Research, were initiated in order to conduct large-scale field studies of organizational structure, organizational climate, environmental factors, job attitudes, individual and organizational effectiveness, illness rates, and personnel retention (Gunderson and Sells, 1975). These studies involved the measurement of individual and organizational characteristics in a sample of 23 naval organizations (20 combat ships and 3 shore stations) and 3 civilian organizations. Major data collection consisted of questionnaires, interviews, on-site observations, medical and personnel data, and organizational records gathered by research teams aboard ships under operational conditions in both the Atlantic and Pacific fleets and at selected shore facilities. The ships included three destroyers, six missile destroyers, three missile frigates, six destroyer escorts, and two aircraft carriers. The principal test instrument, a 400-item questionnaire, included 85 rating scales used to assess perceptions of living and working environments aboard ship (LaRocco, Gunderson, Dean, James, Jones, and Sells, 1975). Illness rates were compared on individual cards for all crew members during 7- to 8-month overseas deployments.

Environmental Stresses Aboard Ship

The human organism in any environment is exposed to a variety of natural and man-made energy sources which may affect psychological and physiological states. There is a vast literature pertaining to physiological and behavioral effects of various biophysical agents. Only a brief review of certain environmental effects relevant to shipboard living is possible here.

NOISE

One advantage of the days of sail was that wind propulsion was generally quiet. Shipboard engines brought high levels of noise with speech interference, distraction and fatigue, and even hearing loss. Noise intensities aboard modern ships are such that noise-induced hearing loss is considered a serious health hazard for certain occupational groups, notably engine room personnel and jet aircraft handlers. The psychological and physiological effects of lesser but pervasive noise sources incidental to crowding and lack of privacy have not been evaluated but presumably take some toll in interfering with concentration, communication, and sleep.

In one study of noise in living spaces aboard a combat stores ship, noise levels varied from 50 to 80 decibels (Webster, 1975). Typical and average values were consistently 57 decibels (A–weighted values) on the upper decks and 67 decibels on the lower decks. These levels were 10 to 20 decibels higher than levels measured in the homes of crew members ashore, yet fewer than 10% of the officers and enlisted men aboard thought the levels unacceptable, and 75% thought that normal speech and work were not affected. Noise levels in the enlisted men's spaces were 10 decibels higher than in the chiefs' and officers' spaces; the average enlisted man also showed greater response in terms of whether the space was uncomfortable or unacceptable and whether sleep, speech, or work were affected.

Webster (1975) reported that noise at a level of 70 decibels is probably acceptable to crew members in work spaces aboard ship. At this level, however, it is almost impossible to hear normal speech sounds at distances of more than a few feet, and certain social activities, such as wardroom conferences or church services, are impossible. Noise in the engine room of the combat stores ship remained at a fairly consistent 87 to 95 decibels. Engine room personnel frequently reported the noise to be "annoying," "very annoying," and "loud." Enginemen generally did not wear protective hearing equipment because they indicated that, by wearing them, they could not hear engine sounds clearly and, therefore, could not do their jobs properly.

Compared to spaces aboard aircraft carriers, offices aboard combat stores ships are relatively quiet, but, compared to offices ashore, they are noisy. People are apparently affected more by noise in their working areas than by noise in their living areas; at least ratings by crew members indicated less tolerance of high noise levels in their working environment.

There is considerable evidence that people become habituated to noise up to levels of 75 decibels. For example, on the combat stores ship persons working in the supply office (67 decibels) found the noise level less tolerable than men working in the engine room (90 decibels) even though noise levels were much lower.

The most sensitive measure of adverse reactions to noise in these studies was its effect upon speech; work problems and discomfort were next in importance. The unacceptability of a space was not monotonically related to the noise level. The job that a man was to accomplish in a noise situation, especially when it involved extensive verbal communications, was more important in determining his reaction than the actual level of the noise. The nature of the noise itself, its time sequence, its connotative aspects (meaningfulness), and its emotional aspects (for example, rock-and-roll music or country-western music are intolerable to some) were more important than the average noise level (Webster, 1975).

Noise abatement can be achieved to a considerable degree in the original design of ships if major noise sources are recognized and dealt with by standard sound engineering methods.

In current shipboard studies (Pugh and Gunderson, 1975) crew members aboard 20 ships were asked to rate noise in their working areas, berthing compartments, head areas (sanitary facilities), messing area, and the ship as a whole on a 5-point scale from "extremely disturbing" to "not bothersome." The individual ratings were summed over the five ship areas to obtain an overall noise score. The mean amount of disturbance attributed to noise varied by type of ship. Crew members from old destroyers of the World War II type reported more severe noise problems than crew members of new destroyer escorts, which were among the most modern ships in the fleet. There was no overlap in mean noise ratings from one ship type to another. Individual scores pooled for all ships did not correlate significantly with number of illnesses during a 7- to 8-month overseas deployment. However, when mean scores for divisions or work groups rather than individual scores were used to predict illness, a significant correlation was obtained ($p < .001$). This result indicated that variations in noise from one work area to another were correlated with differences in illness incidence rates from one work area to another.

HEAT STRESS

Hot environments are known to produce important physiological changes, largely because of a great increase in the body's cutaneous vascular bed during severe heat exposures. Skilled performance has been shown to deteriorate at levels of heat stress that result in a rise in body temperature above normal levels.

Sailors aboard ship, except for engine room personnel, are likely to experience only mild degrees of heat stress, depending upon the efficiency of ventilation and air-conditioning systems. Engine room personnel may be exposed to continuous temperatures in the range of 110° to 140° F, except in spaces directly under the blowers which bring in cool air.

In the study of environmental conditions on 20 ships a large proportion of men aboard the old type of destroyers reported moderate discomfort from heat and poor ventilation, while crew members of modern destroyer escorts were more comfortable. There was some variability in mean temperature-ventilation ratings among ships within type, however, that was not explained by differences in ventilation and air-conditioning equipment. Further analysis of heat stress in various areas within individual ships is needed to identify principal sources of variance.

Individual ratings of temperature-ventilation in work areas correlated significantly with total illnesses during 7-month deployment ($p < .001$), but similar ratings in berthing compartments did not correlate significantly with illness. Individual ratings of the entire ship also correlated with illness ($p < .001$). Mean temperature-ventilation ratings of work areas for divisions rather than individuals also predicted illness ($p < .01$), indicating that variations in heat stress from one work area to another were associated with differences in illness rates.

CROWDING AND PRIVACY AND OTHER CONDITIONS

Ships are densely populated habitats, offering much less personal space per inhabitant than the most crowded slum tenements. The concept of crowding involves more than spatial density; it also involves perceptions of physical and social constraints and affective responses to experienced spatial and behavioral restrictions (Stokols, 1972). Meaningful assessment of crowding requires recognition of all these elements.

Privacy in one sense is merely an absence of crowding, but as a psychological concept it is more than this. Privacy implies definite allocations of space for exclusive personal and private use with secure boundaries and territorial rights. None of these conditions is present aboard ship for the enlisted man, so no real privacy exists for him. "Privacy panels" partially enclosing his bunk provide some semblance of personal space, although these barriers do not prevent unwanted intrusions of sound, light, odor, or work and social activities. Privacy panels apparently make a difference in perceived privacy. In the study of 20 ships, modern destroyer escorts with privacy panels on bunks were consistently perceived by crew members as having more privacy than World War II destroyers with open bunks. Individual privacy scores ("none" to "plenty") did not correlate with illness aboard ship, nor did mean scores for groups of individuals.

Individual crowding scores did not predict individual illnesses. Mean ratings of work area crowding by berthing compartment groups correlated *negatively* with illness. This unexpected result was attributed to the fact that less crowded work areas, such as the engine room (which had relatively high illness and injury rates), tended to be hotter, dirtier, noisier, and more hazardous than other work areas.

Crew members' perceptions of cleanliness ("dirty" to "clean"), odor ("unpleasant" to "pleasant"), safety ("hazardous" to "safe"), color ("unpleasant" to "pleasant"), and lighting ("too dim" to "too bright") also were obtained. Responses on most ships were heavily skewed toward "dirty," "hazardous," "unpleasant," and "too dim." Variations in these habitability characteristics tended to parallel those for noise, temperature-ventilation, and crowding-privacy, in that the old destroyers were perceived unfavorably and the new destroyer escorts somewhat more favorably. Individual cleanliness-odor and safety ratings of work areas correlated with illness in the expected direction, and correlations increased when mean division scores were used instead of individual scores.

Other perceived habitability conditions–food facilities and services, ship services (laundry and ship's store), recreational facilities, and sleep problems (owing to environmental conditions) –correlated in the expected direction with illness and injury rates.

Overall, the most notable environmental predictors of illness rate aboard ship were work noise (mean division scores), work safety (mean division scores), work cleanliness and odor (mean division scores), food facilities and services (mean berthing compartment scores), and recreation (mean berthing compartment scores). Thus, working conditions and personnel services appeared to be the situational factors most clearly associated with differences in illness rates.

The results of this study indicated that ratings of environmental conditions can be used to predict illness and that the predictive validities of these ratings are linked to the actual physical characteristics of the situation. These environmental rating scales, when entered into a stepwise regression equation, yielded a multiple correlation of .41 with illness rate. This situational component of illness variance was independent of the contributions of individual demographic and biographical variables examined in previous research (Rahe et al., 1972; Pugh and Gunderson, 1975).

Navy Adjustment and Reenlistment

Retention of qualified personnel is a continuing issue of major importance in the armed forces. In the study of 20 ships (LaRocco and Gunderson, 1975) sailors completing first enlistments were divided into three categories: those who were qualified to reenlist and did (RR), those who were qualified to reenlist but did not (RN), and those who were not qualified to reenlist (N). The last group was not recommended for reenlistment by commanding officers or was separated from the Navy prematurely. The most important group for purposes of improving retention rates is the middle group (RN), particularly those members who report general satisfaction with the Navy.

Stepwise linear discriminant analysis was applied to the problem of generating a set of functions which could be used to distribute subjects among the three groups. Predictor variables consisted of several biographical items (age, marital status, school performance, and socioeconomic status), a number of service history items (pay grade, job specialty, number of men supervised, type of ship, months spent at sea, etc.), five job attitude measures (Porter's Need Satisfaction Scale, Lawler and Hall's Job Involvement and Intrinsic Motivation Scales, the Overall Navy Satisfaction Scale, and the Habitability Satisfaction Scale, which was based upon ratings of living and working conditions), and a set of service history measures called "Performance Satisfactoriness" (Quarterly Performance Marks, Rate of Promotion, and Demotions). The predictor variables were entered into the analysis in a stepwise manner, and weighted scores were combined to generate discriminating functions which separated the groups by maximizing the differences among them.

The discriminant analysis selected eight predictor variables which accounted for approximately 35% of the variance in the criterion. The group not recommended for reenlistment or discharged prematurely was the most easily distinguished. Its members received the lowest performance marks, the fewest promotions, and the most demotions; they were the most dissatisfied with their jobs, habitability conditions, and the Navy generally; they reported being expelled from school more often; and they comprised the youngest of the three groups.

Distinguishing the other two groups was more difficult. Those who reenlisted (RR) received the highest performance marks, reported the most satisfaction with the Navy, and had the most dependents. Those recommended for reenlistment who did not reenlist (RN) received moderately high performance marks, reported a moderate degree of satisfaction with the Navy, and spent the most time at sea.

A clearer picture of the differences between the groups can be obtained by considering all variables that discriminated regardless of whether they contributed to the discriminant function. Poor school performance items sharply differentiated the N group from the other two. Family income–an indicator of socioeconomic status–distinguished the RR group from the others; those who reenlisted came from lower-income families. Marital status also was a good discriminator of the RR group–a higher proportion of this group was married and had dependents.

Those not recommended came disporportionately from the deck and engineering (blue-collar) divisions. Those who reenlisted were least likely to come from these job types. Reenlistees were most likely to have received technical training (A or B school); those not recommended, the least likely. Mean General Classification Test scores were highest for the RN group.

The measures of "satisfactoriness" contributed greatly to the discriminant functions. One quarterly performance mark, Military Behavior, was among the most important variables, and the three remaining performance marks, Professional Performance, Military Appearance, and Adaptability, also discriminated in the expected direction. One index of satisfaction (Overall Navy Satisfaction) entered into the discriminant function, but all others were significant as well. Three habitability satisfaction subscores (work, berthing compartment, and entire ship), Job Involvement, Intrinsic Motivation, and Need Satisfaction were all high for the reenlistees, moderate for those recommended for reenlistment who chose to leave, and low for those not recommended.

The group profiles derived in this study were similar to profiles reported in other retention studies (Federman et al., 1973). Discriminant weights of the type developed here could be used to assess reenlistment potential at critical points during a man's first enlistment; the resulting probabilities would be useful for career counseling, making special training assignments, and taking other personnel actions for which career potential is an important consideration.

Summary

Health and adjustment in naval organizations are complex functions of individual characteristics, environmental and organizational influences, and the interactions of these sets of variables. Current studies of men aboard ship are concerned with the development of a social-system model to represent behavior in naval organizations and the measurement of physical environments of ships, habitability perceptions of crews, organizational structure, organizational climate, leadership patterns, biographical and military service information, job motivation and satisfaction, and medical and performance criteria at the individual and organizational levels. Preliminary findings indicate that shipboard environmental conditions are correlated with health, satisfaction, performance, and retention of naval personnel.

BIOGRAPHICAL CONSTRUCTS AS PREDICTORS OF ADJUSTMENT TO ORGANIZATIONAL ENVIRONMENTS

Paul D. Nelson

For most individuals, adaptation to the varied demands of military service represents but another set of obstacles to be hurdled in the track-of-life experience. For some, however, the demands are too great or the coping mechanisms inadequate for effective service to be rendered. Institutional parallels can be drawn, of course, with the school years, marriage, civil employment, and even retirement. But the military organizational environment is perhaps more restrictive in behavioral degrees of freedom tolerated with its institutional prescriptions of discipline and 24-hour readiness in the performance of duty.

Psychiatric casualties among military personnel in times of peace as well as war have been amply documented manifestations of inadequate adjustment to the institutional and situational demands of duty (Arthur, 1966; Ginzberg, 1959; Haggard, 1949). Still other instances of misbehavior or immaturity reflecting adjustment difficulties result in administrative sanctions. And though the percentage of military personnel failing in such manner to adjust may be relatively small, the associated costs, to individual and organization alike, have been sufficiently great to warrant considerable attention. As Rioch (1968) stated, the prevention of such maladjustment has become a major task for military psychiatry.

To predict and prevent maladjustment to military service requires an understanding not only of the individual and of the environmental demands imposed but of the complex transactions relating the two. Concepts of psychological stress and coping become important in this context when viewed in terms of dynamic transactional processes, as suggested, for example, in the writings of Lazarus. The very process of coping, in the view of Lazarus and his colleagues (1974), entails perceptual, cognitive, emotional, and behavioral dispositions of individuals relative to the environment at hand. When

AUTHOR'S NOTE: *Opinions asserted are those of the author and are not intended to reflect the official views or endorsement of the Department of the Navy.*

stress and adaptation are viewed in such a manner, as resultants of natural and continuous person-environment transactions, it becomes as important to understand how individuals succeed as it is to know how they fail.

What course might we take, then, as social psychologists and other students of human behavior, in assessing and predicting adjustment to the organizational environment of military service? There are many possibilities; but I suggest as one course that we attend more closely to human biography, the life histories of individuals which reflect in a continuous trail the processes and outcomes of coping experience. Biography, of course, has long served as an important source for the development of hypotheses, if not theory, about the structure and function of human personality. In particular, the school of humanistic psychology has been dedicated to understanding the course of human life as a unified whole, implying a knowledge of life history.

Biographical data should serve as a bridge in the analysis of human behavior between the demographic, cultural, and social-institutional domains of the sociologist and the human abilities, attitudes, and personality domains of the psychologist. In their most complete form, such data should embrace the characteristics of human environments as well as those of the individuals and their behaviors, both achievements and failures, in those environments. On all such accounts, the social psychologist has challenging territory to explore.

Methodological issues in life history research are not trivial, of course—the controversies of ideographic and nomothetic models being among the central issues. In his discussion of the psychology of population, Back (1973) advances some interesting notions along conceptual and methodological lines, referring to individual lives as a series of events and decisions adding up, in summary, to the demographic trends in society. The events of life studied by the demographer, Back argues—namely, those pertaining to family, schooling, occupational decisions, geographic and social mobility, and other vital statistics of life processes—are typically those of great significance in the fate of human lives. Consequently, they serve as potential standards of reference or checkpoints across the lives of individuals. What are the psychologically meaningful features of behavior for individuals across demographic events? Back asks; and how consistent or variable are individuals across situations of life, even those transpiring over decades? How predictable is the individual through demographic events? Do individuals vary in the amount of control that they exert over the course of their lives?

For the most part, even the simplest of such conceptualizations have been typically absent, by contrast, in the use of biographical data for personnel assessment in industry or the military (Nelson, 1971)–settings in which life history data have nevertheless been related to adjustment and performance. The predominant emphasis in such studies has been more empirical than theoretical, more predictive than explanatory. Biographical variables useful in some situations seem less so in others, but relatively little effort seems to be devoted to understanding why.

Against that background some recent trends in research do focus more on the development of constructs than on the predictive validity of discrete items of life history. By way of example, W. Owens (1971) has attempted dimensionalizing life history data in the form of typologies or biodata profiles reflecting behavioral dispositions. From a more ecological point of view, the increased attention given in recent years to describing and measuring environmental characteristics is promising (Insel and Moos, 1974). Barker (1968), in particular, has conceptualized the significance of behavior settings in accounting for consistencies and variability of an individual's behavior across time and environmental contexts. The theoretical and methodological issues addressed in such work are vital to advancing transactional perspective in human biography.

In his analysis of stability and change in human characteristics, Bloom (1964) argues, as others do, for more adequate measurement of the environmental context in which behavior and growth occur but also for the importance of longitudinal research, certainly a central feature of biographical data. Such an orientation is nicely illustrated in the monumental work of Bachman and his associates (1967) in their appraisal of youth in transition, from adolescence to young adulthood, from school to work environments and roles.

When panel survey research methods are used, longitudinal and implicitly biographical research on the "fit" between young men and their environments reveals over time the complex and contingent relationships of individuals' behaviors, attitudes, and aspirations with the family, peer group, community, school, and work environment. Like the classic cross-sectional studies of adolescent society by Coleman (1961) and of behavior in big schools and small schools by Barker and Gump (1964), the research of Bachman and his colleagues is especially important as background for military service adjustment. The adolescent years, and the many adjustments therein, are source for the significant demographic events, institutional networks, and stage-of-life transitions immediately preceding military service for most individuals in our society.

Initial focus in the Bachman studies has been upon the high school adjustment of boys from different walks of life, with attention given first to the influence of family background and pupil intelligence (Bachman, 1970), leading next to the prediction and interpretation of "dropping out" behaviors (Bachman et al., 1971). Analysis has also been made of the disposition of young men toward military service relative to other endeavors following high school (J. Johnston and Bachman, 1972). At that point the fate of young men actually entering military service becomes an object of research in the armed forces themselves.

Indeed there is continuity from the Bachman studies to those of military adjustment. Against criteria of psychiatric consequence, disciplinary problems, and overall effectiveness during first enlistment, the age at which young men enter service, their school behavior history, and the general aptitude of enlistees have been observed to be quite valid predictors of military adjustment (Plag and Hardacre, 1964). The interactions among those variables and others, however, should be noted in terms of their significance for the development of biographical constructs.

General aptitude and the level of schooling attained appear in all services to be favorable indicators of military adjustment during first enlistment—in rather linear fashion, at that, when analyzed separately. This may, of course, be as much a function of the duty assignments associated with such attributes as the character of the individuals themselves. The age at which young men enlist, however, appears to serve more as a moderator of aptitude and schooling, interacting with those two variables to contribute uniquely to prediction of adjustment. What does the age of enlistment reflect in itself, in relation to when the individual left school, how far he went in school, and his general aptitude? Perhaps motivation is tapped; perhaps situational contingencies of further education, of employment opportunities, or of family circumstances are also involved. In terms of actual service experience, what is the significance of being an older or younger recruit? Such questions have not been altogether answered.

In line with Bachman's discussion of the different reasons why boys leave school prematurely, the studies of naval recruits conducted during the past decade reveal school expulsions or suspensions to be especially powerful predictors of maladjustment during first enlistment (Arthur, 1971). Such reflections of previous difficulty in adjusting to the rules, regulations, or other conventional standards of behavior also modify the predictive validity of aptitude and educational attainment against service adjustment criteria. An absence of such disciplinary history can render a young man with lower educational achievements and aptitude a better risk for service adjustment than a seemingly brighter and more educated peer who did not get along with school authorities to the point of having been expelled or suspended.

Among the psychiatric problems encountered in military service, personality disorders of assorted nature are most prevalent. That construct in psychiatric nosology can certainly account for a fair portion of the recruits having a history of disciplinary problems, but, as Wilkins (1961) indicates, it is one of the most diagnostically difficult for personnel screening. But life history data of the sort which portray person-environment transactions must certainly be considered among the better sources of predictive information for such screening tasks.

Other biographical data pertaining to earlier life experiences in family and community settings, including interests and avocational pursuits, have also been observed in relation to criteria of military service adjustment (Gunderson, 1963). Their significance, however, as somewhat similarly observed by Bachman and his associates, is typically reflected in school adjustment, serving more for interpretation of behavior in that setting than as unique predictors of subsequent military service adjustment, assuming that school behaviors have already been accounted for. Exceptions may pertain to biographical constructs developed with specific reference to a particular type of behavior, as illustrated by recent research on drug abuse. While preservice indicators of general adjustment tend to differentiate drug abusers from others during first enlistment in military service (Plag and Goffman, 1973) and while histories of family, school, and community adjustment difficulties also appear related to preservice drug abuse (Kolb et al., 1974), though complexly so (L. Johnston, 1973), the development of a drug-involvement scale (Gunderson et al., 1973) based upon a detailed preservice history of drug-related behavior patterns may offer promise for predicting similar patterns of behavior during military service beyond what can be predicted from the more general items of preservice life (Kolb et al., 1975).

All such research reveals an important facet of human biography, namely, its temporal nature. The longitudinal perspectives of Bachman (1970) and Bloom (1964) lead those authors to suggest that the greater the chronological proximity of predictor to criterion measures, in general, the greater their relationship is expected to be, assuming any conceptual linkage of the characteristics or behaviors at issue. Such a postulate seems reasonable enough, perhaps obvious. Yet in the use of biographical data to predict behavior it often seems overlooked.

In support of that postulate, let us return for illustration to the longitudinal study of Navy recruits during their first enlistment (Plag and Goffman, 1966). At the time of enlistment, of course, only preservice histories are available from which to forecast the probability of effective service during a 4-year enlistment. How the individual handles the stresses of recruit training adds further information on adjustment potential, and how one performs in one's first duty assignment adds still more. By the end of the first 2 years of service, if a young man is still on duty, quite a different set of variables predict effectiveness for the final 2 years, more accurately too, than do so at the time of enlistment or even by the end of recruit training. In essence, service behaviors themselves become better predictors of future service behaviors than are preservice behaviors.

In a parallel study of first-enlistment Marines, much the same was observed in predicting combat performance among young men who at the time of combat were engaged in their third and fourth years of military duty (Mahan and Clum, 1971). Similarly, though in quite a different setting of isolated Antarctic station groups, preservice behaviors were related more to the performance of younger Navy men than their more senior station mates, whose adjustment, in turn, was more predictable from knowledge of their previous histories of military service (Nelson and Gunderson, 1963). And, in another example, for naval personnel who become psychiatric casualties after some time in service, the more recent service histories appear better predictors of posthospitalization adjustment than are preservice histories (Edwards and Berry, 1974).

In addition to their relationship with social adjustment, the temporal elements of life history also seem important in health, and critical research has been conducted in recent years on psychosocial factors in the etiology of health change (Gunderson and Rahe, 1974; French et al., 1974; Holmes and Rahe, 1967; Levi, 1971). Illnesses, like the changes-in-life circumstances which seemingly both precede and follow them, tend to be clustered in the lives of individuals. Rahe's analysis (1972) of changes in one's family, work, and economic and social spheres of life among military as well as civilian populations points again to the recency of experience reflected in life histories as significant in predicting near-future illness episodes. More needs to be known about the role of the individual relative to such life changes, the coping processes called forth, and how these in turn are related to physiological pathogenesis (Nelson, 1974); but the development of biographical constructs for predicting illness seems important in a society and an era in which the frequency of change and the pace of life are indeed great and complex.

A part of the temporal feature of life histories—and perhaps the most perplexing of problems in biographical research—is that of life's continuities and discontinuities. For while many individuals seemingly persist in their previous patterns of behavior or health, for better or worse, others outlive their past, and still others fail to live up to it. Therein, of course, lies in good part the source of our inability to predict future adjustment from that of the past with little more than modesty. Therein also lies, it would seem, our greatest challenge as students of human behavior—not only to predict but perhaps also to facilitate more effective adjustment to organizational environments.

In the studies cited of naval service adjustment, even among the initially poorest risks of recruits, close to half or better can be expected to serve effectively during first enlistment, many of them in a sense turning their lives around. Similarly, among those returned to full duty following psychiatric treatment, about half can be expected to serve effectively. And, among those whose recent life history is replete with turmoil and change and even minor illness, not all get sick in the near future. What characteristics of the individuals and their environments, what transactions between the two, can we therefore identify to differentiate these outcomes better? Can we do this at least in part through the development of biographical constructs?

Certainly social psychologists will not be alone in pursuit of such answers. But their longitudinal mapping of person-environment "fits" as pursued in the works of Bachman, French, Rahe, and their colleagues, as well as the paper by Gunderson in this volume, are efforts in the right direction. And though such efforts are not necessarily couched in terms of biographical research, their longitudinal perspective and the types of data gathered render them very much related to the concept of human biography. For what constitutes the environmental context and behavior of today becomes in a sense the life history source of biographical data tomorrow, not altogether independent of that of the past.

Moos (1974) summarizes the value of biographical data primarily in terms of their unobtrusive and global validity for predicting behavior but questions their utility as yet for advancing our understanding of coping and adaptation to stress. I would add that, in concept, life history may afford the social psychologist as much opportunity as any source of behavioral data from which to advance our understanding of such phenomena. The potential lies in the development of biographical constructs which, as Radloff and Helmreich put it (1968:198), represent "the psychological significance of essentially sociological data." The organizational environments of military service furthermore afford the social psychologist a particular setting, and an opportunity, to further develop and test such constructs with the goal of improving the quality of life for individuals during a significant part of their maturational and productive years.

WORK-RELATED ATTITUDES OF MILITARY PERSONNEL

David G. Bowers

Under an all-volunteer system the armed forces must compete in the manpower market. Like other types of employment, military service must provide work roles which are satisfying activities in their own right, which are seen as making a positive social contribution, and which provide adequate financial rewards, fringe benefits, and the like.

To the casual observer, as to the social scientist, it appears that conditions which have existed since the start of World War II may be shifting. Many of the tenets, assumptions, and customary relationships of the last three decades, some forming the basis for military manning and management practices, are undergoing great changes. In many ways affluence has rendered meaningless a number of the accustomed motivational strategies which were effective in the past. Attitudes toward authority and toward the value of great openness, questioning, and candor all appear to be changing. Not only the military services but most of the major institutions of our society as well seem to be faced with the necessity of closely examining, and perhaps greatly altering, practices based upon old assumptions in these areas.

The research which this report summarizes began with the proposition that changing values, expectations, life styles, and preferences for the quality of organizational life are important and perhaps overriding considerations in relation to the fortunes of an all-volunteer force. This proposition stems from two sources:

(1) Accumulating data of a formal variety suggest that in recent years noneconomic matters have become increasingly central to an ever larger number of persons.

(2) In a great number of instances, increasing in frequency, there have been dramatic shifts in the behavior of persons and organizations on dimensions related to value and quality-of-life issues.

AUTHOR'S NOTE: *The research upon which this article is based was supported by Office of Naval Research contract N00014-67-A-01810048. The cooperation and help of that agency is hereby gratefully acknowledged.*

Although one may reasonably question the extent to which an affective or emotional change in attitudes has occurred over the years, there appears ample ground for assuming that the informational and behavioral components of attitudes have changed markedly. Today's likes, dislikes, and preferences may be little different from those of two generations ago, but they are supported by a much sturdier informational substructure, and the behavioral repertoire in which they are seen as potentially finding expression contains a much wider array of alternatives, few of them in the category of "compliance." It may well be, in other words, that values themselves have changed less than have certain other things associated with those values, like willingness to tolerate practices at odds with them, perceived available alternatives, ways of behaving in response to disliked practices, and so on.

Much, therefore, hinges upon the acceptability and up-to-date character of military manpower practices, since it seems likely that little by way of socialization (attitude change of servicemen in directions more compatible with customary services practices) can be expected. Unfortunately, the degree of such correspondence seems lower than what could be desired. Although alternatives have undergone vast change and expansion since the early years of this century, managerial practices have changed relatively little.

Method

SUBJECTS

To answer these and other questions, a survey was administered to a sample of Navy units and to a national random sample of civilians. A detailed description of the sampling techniques as well as a description of the fit of the samples to their respective populations is presented in a technical report, *A Methodology for the Studies of the Impact of Organizational Values, Preferences, and Practices on the All-Volunteer Navy* (Michaelsen, 1973).

NAVY SAMPLE

Data from the total sample of 2,522 Navy personnel were collected from both ship and shore stations (38 different sites) between November 1972 and February 1973. The surveys were personally administered by staff members from the Institute for Social Research.

Personnel actually surveyed at a particular site are members of intact organizational subunits, consisting of work groups related to one another through supervisors who are, at the same time, superior in the group that they supervise and subordinate in the group immediately above. In this fashion, one may conceive of the organization as a structure of such overlapping groups, a pyramid of interlaced pyramids. For purposes of identifying and selecting intact units for the study's analytic aims, the sampling basis was designated as a "module," which means a "pyramid" of groups three echelons tall. Thus, members from four adjacent levels were included, with the module head defined as the person at the apex of that particular three-tier pyramid. Yet another criterion for the selection of a module was that the person at the apex (the module head) had been at his current assignment for at least 3 months.

An appropriate number of module heads were randomly selected from a list of all personnel at a site who met the criteria for module head. If a particular module did not provide a large enough sample of personnel required for the particular site, another module head was selected by the same method. Thus, the sample from a site consisted of one or more modules.

CIVILIAN SAMPLE

The civilian data collection was conducted during February and March of 1973, as part of a larger interview study conducted by the Survey Research Center. The sample included 1,327 dwelling units selected by a multistage sampling system so as to be representative of all dwellings in the conterminous United States exclusive of those on military reservations.

At each housing unit an interview was conducted with a specific designated respondent, male or female, 18 or older. The final segment of the interview consisted of questions related to the all-volunteer force. After this personal interview, respondents were asked to complete a written questionnaire (the civilian version of the ONR instrument). In addition, copies of the questionnaire were administered to a supplementary sample consisting of all other individuals age 16 or older who were present in each household at the time the interview was taken. A total of about 1,855 civilian questionnaires were provided. We have chosen to treat both samples as a single, unweighted sample of people age 16 and older throughout the United States, since there were no systematic differences between the two.

MEASURES USED IN THE PRESENT ANALYSIS

The measures employed in this present analysis are described in considerable detail in the separate methodological technical report (Michaelsen, 1973). It seems sufficient for present purposes merely to list them categorically:

Values and Preferences

Preferred Leadership Measures–8 indices, tapping the extent to which the respondent desires particular forms of behavior from his supervisor and his coworkers (peers).

Job Characteristics–16 measures of the desirability or importance of certain properties in one's ideal job.

Management Beliefs–adherence by the respondent to authoritarian beliefs concerning effective management styles and his awareness of the importance of human factors.

Actual Practices

Measures paired to each of the Preferred Leadership and Job Characteristic measures, asking respondents to describe how it actually is at present in their current work roles.

Demographic Variables

Of a number of possibilities, a more limited array of self-reported background characteristics are selected for the analyses reported herein: Sex, Age, Education, Region of Origin, Community of Origin (rural-urban rating), Race, Reenlistment Intention, Rank.

Results

VALUES AND PREFERENCES IN THE WORK SETTING

In order to search for value differences of the kind described, emerging in the American population generally and potentially affecting the necessary manpower practices of the Navy, the responses of all persons in both the civilian and Navy samples to value and preference measures were stratified by six demographic characteristics. Of these, two—age and education—assume principal importance in fact in relation to the issues to be discussed.

AGE-RELATED PREFERENCES

Three subsets of work-life related values and preferences concern us in the present study: (a) preferred characteristics of the *job;* (b) preferences regarding the behavior of one's supervisor and peers (his leadership style and their styles in dealing with one another); and (c) adherence to a set of beliefs which are more or less democratic (as opposed to autocratic).

Our findings would suggest that constancy, rather than difference, is the rule with regard to the first of these preferred *job* characteristics. When the 14 job preference measures were rank-ordered for Navy men and compared to a similar rank ordering for employed civilian men, the two sets of rankings correlated quite

Table 1. MOST AND LEAST IMPORTANT FEATURES OF A PREFERRED JOB*

Overall Rank	Civilians	Navymen
Most Imp. 1	Opportunity to Control Personal Life	Opportunity to Control Personal Life
2	Good Pay	Good Pay
3	Friendly People	Avoiding Bureaucracy
4	Good Fringe Benefits	Good Fringe Benefits
5	Avoiding Bureaucracy Mean = 3.58*	Challenging Work Mean = 3.57
10	Opportunity to Serve My Country	Opportunity to Serve My Country
11	No One to Boss Me	Lots of Free Time
12	Clean Job	No One to Boss Me
13	Lots of Free Time	Prestigious Job
Least 14 Imp.	Prestigious Job Mean = 2.52	Clean Job Mean = 2.58

*Importance Scale: 1 = Very Unimportant; 2 = Fairly Unimportant; 3 = Fairly Important; 4 = Very Important.

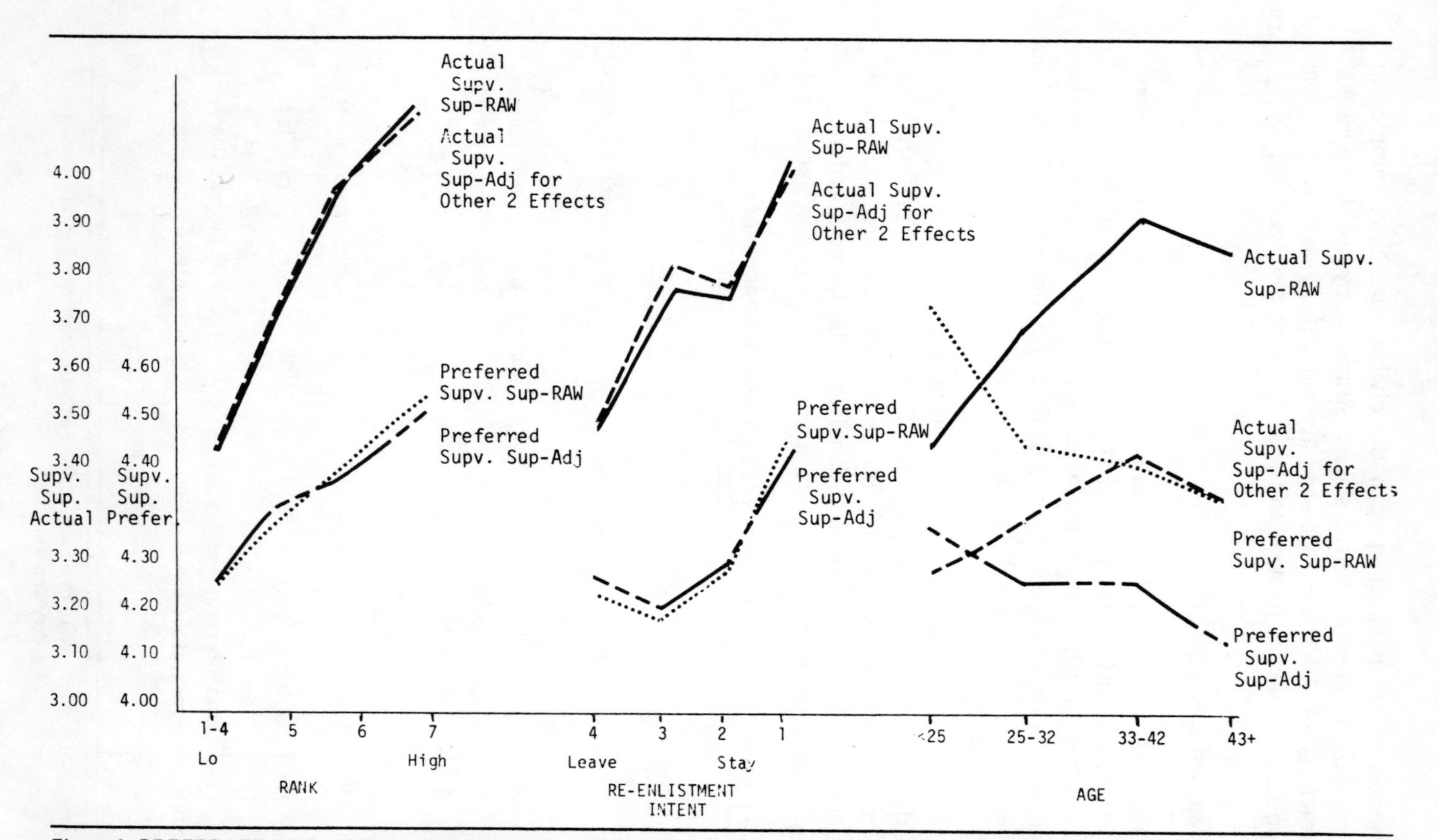

Figure 1: PREFERRED AND ACTUAL SUPERVISORY SUPPORT, BY RANK, REENLISTMENT INTENT, AND AGE

highly (.90). Even among age groupings of civilians, the relative rankings were very much the same (average correlation = .90).

As the data in Table 1 indicate, both Navy men and civilians attach the greatest importance to personal independence (controlling one's personal life and avoiding entangling bureaucracy) and to economic success (good pay and fringe benefits). The job characteristics which *least* concern them are cleanliness, prestige, free time, absence of a "boss," and, perhaps surprisingly, an opportunity to serve one's country.

These findings seem to indicate that, despite the rhetoric of recent years, the traditional American values of independence and material success still are important and are likely to remain so for the immediate future. However, our findings do indicate one set of differences that is particularly striking. Navy men 43 years of age and older, whether enlisted men or officers, present rank-ordered profiles on these preferred job characteristics that are unlike those of (a) young enlisted men, (b) young officers (who closely resemble young enlisted men), or (c) civilians their own age. These dissimilarities occur largely because of the importance attached to the opportunity to control one's personal life (which older Navy men do not value as highly as do others) and service to one's country and challenging work (which older Navy men value *more* highly than do others).

In the area of leadership preferences, a rise with age appeared in the Navy data which does not appear, or appears only slightly, among civilians. These rises in leadership preferences with age appear to reflect the masking effects of rank and self-selection. Figure 1, which shows one of the measures (Supervisory Support), in relation to background variables (Rank, Reenlistment Intention, and Age), is illustrative of a general pattern of findings:

(1) Controlling for other variables has little effect on differences by Rank.

(2) Controlling for other variables has little effect on self-selection (measured in this instance by Reenlistment Intention).

(3) Controlling for other variables removes the effect of Age.

(4) Effects are stronger for Actual than for Preferred leadership.

Although any discussion of cause-and-effect relationships is somewhat speculative for findings that are derived, as these are, from those data collected at a single point in time, the most parsimonious

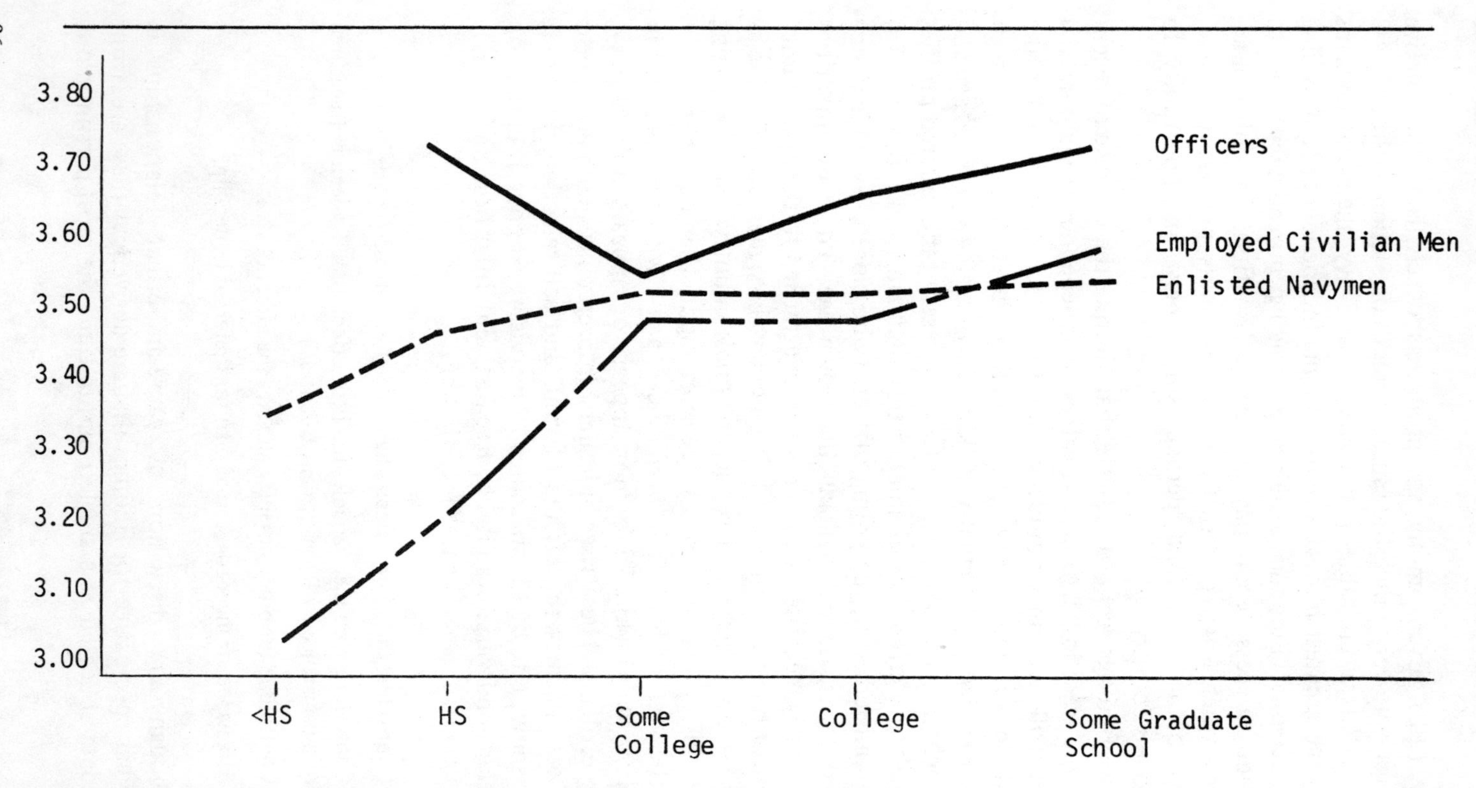

Figure 2: IMPORTANCE OF CHALLENGING WORK BY EDUCATIONAL LEVEL, FOR ENLISTED NAVYMEN, EMPLOYED CIVILIAN MEN, AND OFFICERS

explanation for these results would begin with the behavior actually experienced and move from that to preferences. In descriptive terms, Navy men in any age category report to supervisors whose behavior encompasses a fairly broad range, from quite good to very poor. The *average* behavior experienced rises in positiveness with age, partly because of rank (higher-rank persons are supervised by persons of even higher rank who are, on the average, themselves better supervisors) and partly because of self-selection (specialties, career choices, and assignment practices result in some situational constancy across the period of service, and those who experience comparatively poor situations leave the service). That such effects are more apparent for *actual* than for *preferred* leadership characteristics adds weight to the argument that persons quite naturally are influenced in the setting of their aspirations by their actual experiences.

The third major area—autocratic versus democratic beliefs—will be treated only briefly at this point. In general, there would appear to be a trend toward more autocratic beliefs with age; however, this seems to be intertwined with the effects of educational level. For this reason, further treatment of this topic will be deferred to a subsequent section.

PREFERENCES RELATED TO EDUCATIONAL LEVEL

The findings in relation to education display both consistencies (among job characteristic preferences) and differences (for leadership preferences) when Navy men and employed civilian men are compared.

For both Navy men and civilians, greater education is associated with reduced concern about economic issues, less importance attached to service to one's country, and enhanced concern about having challenging work (see Figure 2). Among Navy men, greater education is also associated with more importance attached to personal independence. Stated thus generally, a number of interesting, though minor, differences are perhaps concealed:

(1) In the economic area, concern about fringe benefits declines with education for enlisted men, for officers, and for civilians. However, whereas the importance of *pay* declines with education for enlisted Navy men, the importance of *steady work* (without layoffs) declines for employed civilians. Neither measure declines for officers.

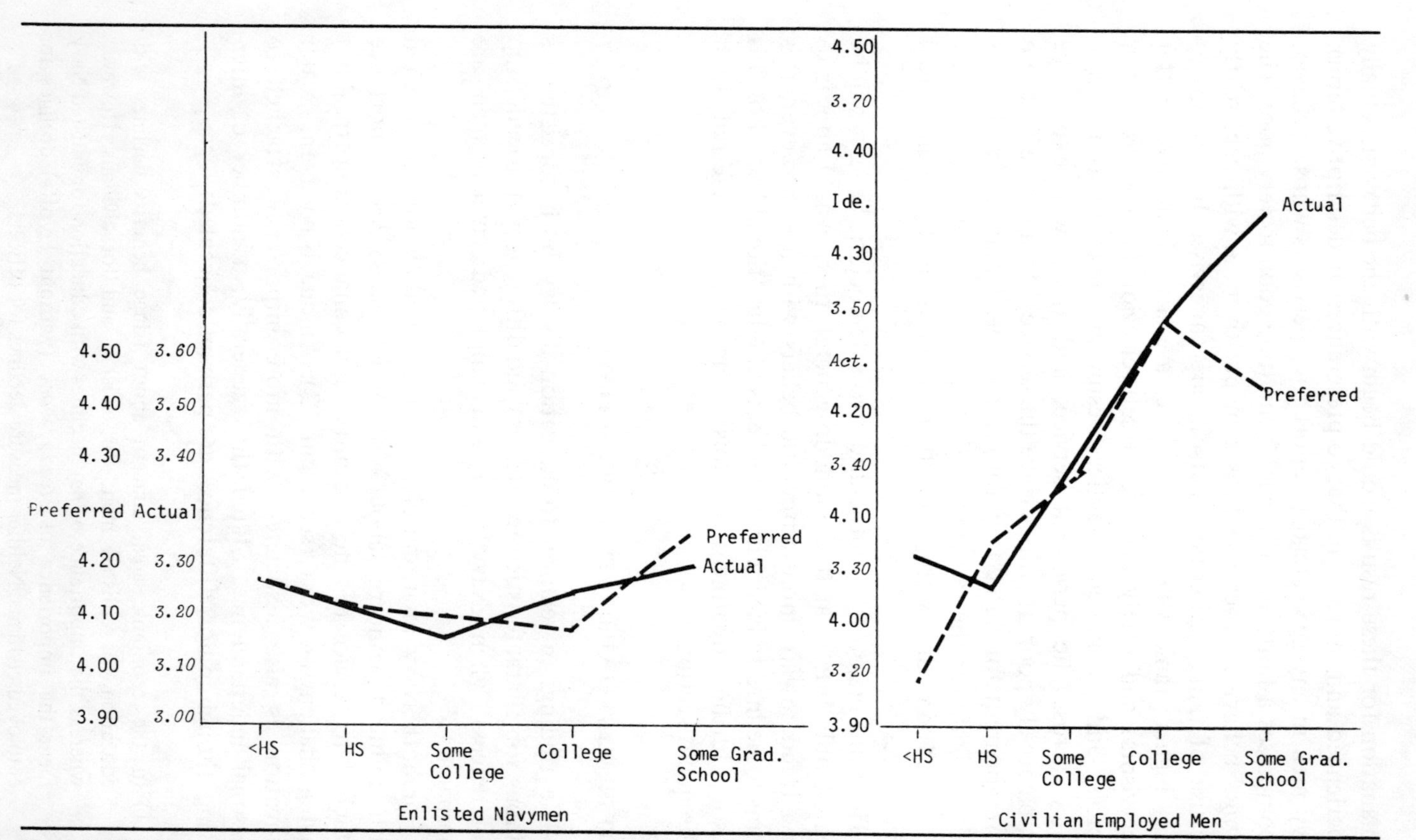

Figure 3: MEAN PREFERRED AND ACTUAL LEADERSHIP OF ENLISTED NAVYMEN AND EMPLOYED CIVILIAN MEN

(2) Much of the steeper rise with education of preference for challenging work among civilians is attributable to the lower end of the education scale (those with a high school education or less), a feature present only slightly in the enlisted Navy men curve, and not present at all for officers.

Turning to leadership style preferences, we see that nearly all of the statistically significant difference among educational categories of Navy men, apparent when the combined sample was considered, disappears when enlisted men and officers are considered separately. It thus appears to reflect the combined effects of (a) difference between these two categories of personnel and (b) the different distributions of these two groups across educational categories (see Figure 3).

Among civilians, however, a definite rise in preferred leadership with education occurs in a form considerably steeper than that for enlisted Navy men. For civilians, as for Navy men, the data rather clearly suggest that leadership preferences are set in some relationship to actual experiences. Although levels of actual and preferred leadership differ, the two curves are in each case similar in shape.

AGE, EDUCATION, AND AUTOCRATIC VERSUS DEMOCRATIC BELIEFS

An objective discussion of the issue indicated in this heading is difficult, largely because of semantic difficulties. Thus, in organizational life, "autocratic" rapidly becomes "authoritarian" and brings to mind sadistic regimes from the history books. In an administrative context, "democratic" similarly changes to "one man, one vote," and from there to notions of disorder and absence of direction.

Nevertheless, there is a dimension of behavior or practice coordinate with a set of beliefs similarly arranged. Toward one direction these behaviors and beliefs become increasingly reliant upon formal authority, more insistent upon artificial distinctions of status and position, more distrustful of the motives and capabilities of others. Toward the opposite direction behaviors and their allied beliefs become less status-conscious, more trustful, and more concerned about persuasive competence, from whatever source.

Although many terms could be applied to these directionally opposite styles, perhaps "domineering" and "cooperative" are most descriptive. In the present study, the general finding is that belief in

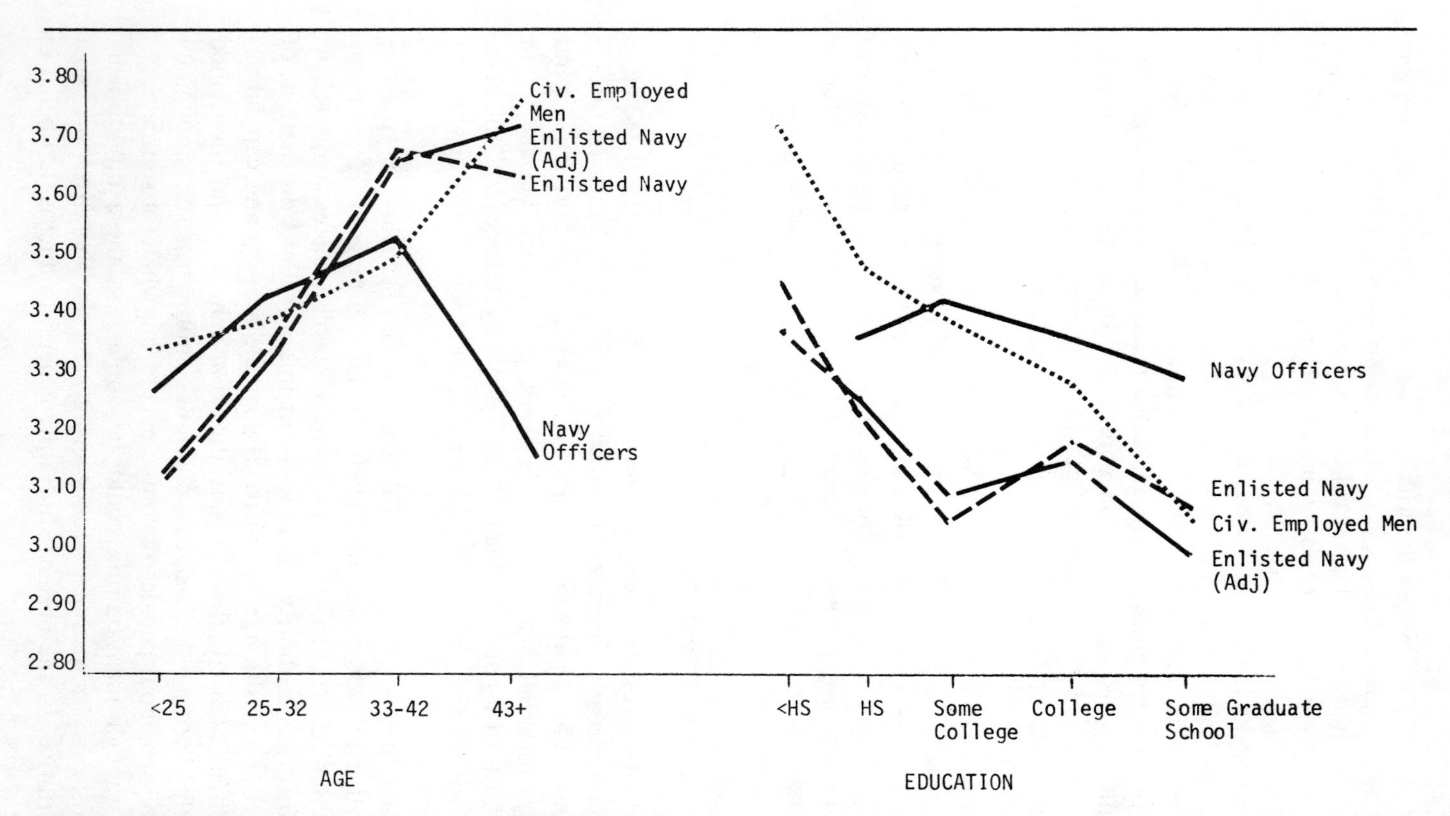

Figure 4: ADHERENCE TO AUTOCRATIC MANAGEMENT BELIEFS

autocratic (domineering) supervisory practices (a) rises with age and (b) declines with education. Figure 4 illustrates this quite clearly, along with certain qualifications:

(1) The curve by age for Navy officers looks remarkably similar to a comparable curve for civilians, rising until age 42; for the highest age category, however, the two curves reflect distinctly different values. Older Navy officers are among the *least* autocratic of groups.

(2) In this fact, older Navy officers seem to resemble young Navy officers, who are distinctly less autocratic than their civilian counterparts.

(3) Controlling the enlisted age curve for the effects of rank, self-selection, and education has little effect. Perhaps the greatest gap along plotted points is that between the youngest enlisted men (mostly first-termers) and the older enlisted men who for the most part supervise them.

(4) Controlling the enlisted education curve for the effects of age, rank, and self-selection has similarly little effect. In general, the decline with increasing education remains.

Practices in the Work Setting

THE NAVY: SHIP AND SHORE

Figure 5 presents graphically, for the total Navy sample and for its ship and shore components, those measures which constitute the critical indices of the *Survey of Organizations.* The measures show at what percentile point on this national array of respondents the mean Navy respondent score falls.[1]

Judging what constitutes being "normal," better than average, or relatively low is an arbitrary, subjective process. In the present instance, we shall establish at the outset the convention of considering the 40 and 60 percentile marks as the boundaries of the normal or "typical" range, with those measures below that range considered potential problem areas, those above it indications of organizational vitality and strength.

As the charted data indicate, on the standard indices of the S.O.O. the Navy in toto falls within the normal range on all but the following measures:

–All measures of organizational climate, but especially Motivational Conditions (for which the Navy respondent is lower than nearly three-fourths of

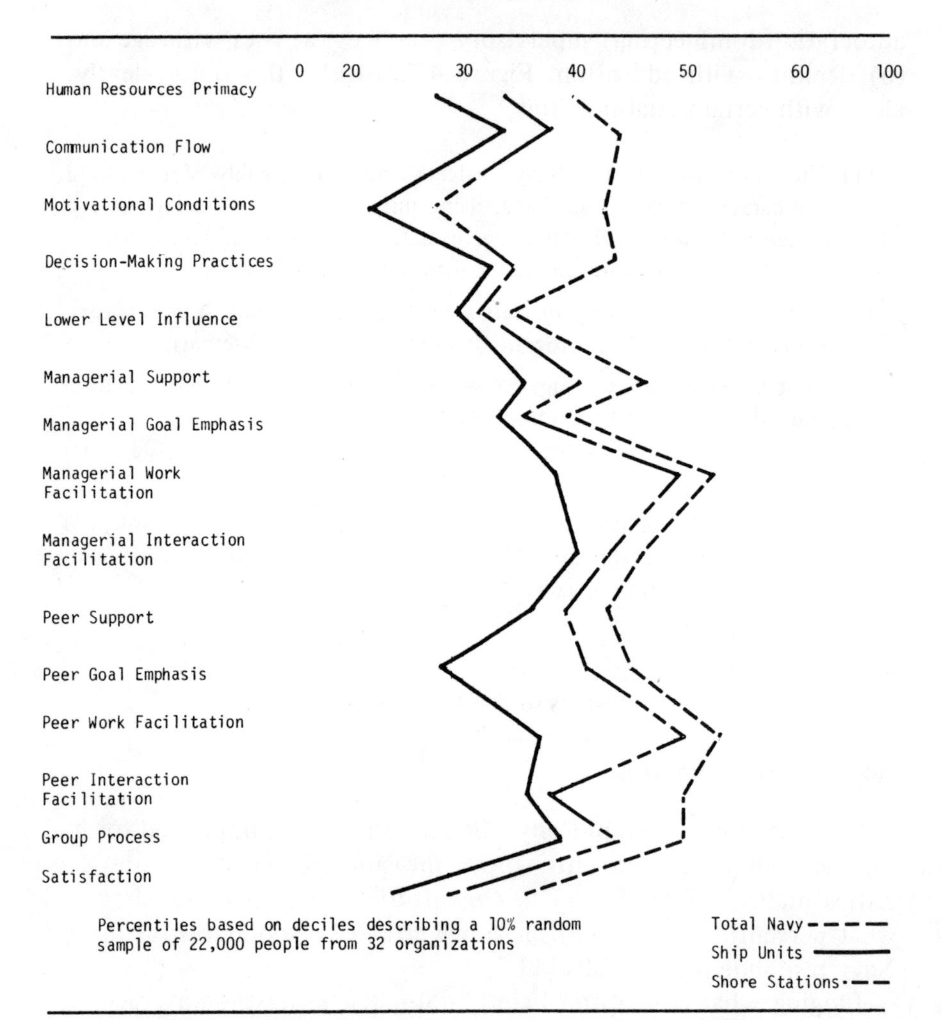

Figure 5: PERCENTILE PROFILE FOR SHIP, SHORE, AND TOTAL NAVY MAJOR S.O.O. INDICES

the civilian industrial respondents); Lower Level Influence (for which he is lower than approximately two-thirds of the civilian respondents); and Human Resources Primacy (lower than two-thirds of the civilian respondents).

–Managerial Goal Emphasis.

–Satisfaction.

Further scrutiny of the items making up these indices indicates that the lowest item scores occur on Satisfaction with the Organization (20th percentile), Conditions Encourage Hard Work (23rd percentile), and Satisfaction with the Job (25th percentile). Taken together, they suggest that the conditions of organizational climate which impinge directly or indirectly on the performance of one's Navy job are seen in a distinctly negative light.

Additional items, not contained in the *Survey of Organizations* standard item list but included within the present questionnaire for other purposes, provide additional insights concerning what it is that Navy respondents do and do not mean when they describe conditions as discouraging and jobs as less than satisfying. The data suggest that there is no appreciable difference between Navy men and civilians in industrial organizations on the following:

–Whether there is or is not someone to boss them in their work.

–Whether their job provides a chance to learn new skills.

–How hard they are required to work.

–How clean their jobs are.

–Whether their job provides a chance to get ahead.

–How much responsibility they must assume.

–How much free time the job permits.

–Whether their job is one whereby they can help make the world a better place.

To this must be added that array of characteristics upon which Navy men describe their jobs as distinctly *different* from those of civilians:

–As one might expect, more civilians feel negatively about their prospects for steady employment than do Navy men.

–More Navy men feel that, although their jobs require that they learn new skills, those jobs do not permit them to use the skills and abilities which they have and gain; and they do not view their jobs as particularly prestigious.

–Although more Navy men than civilians describe their fringe benefits in favorable terms, many more Navy men than civilians view their pay in negative terms.

–Although more Navy men feel that their jobs offer them a chance to serve their country, an even larger proportion feel that it does not allow them to stay in one place (even though, by and large, they are no more attracted to

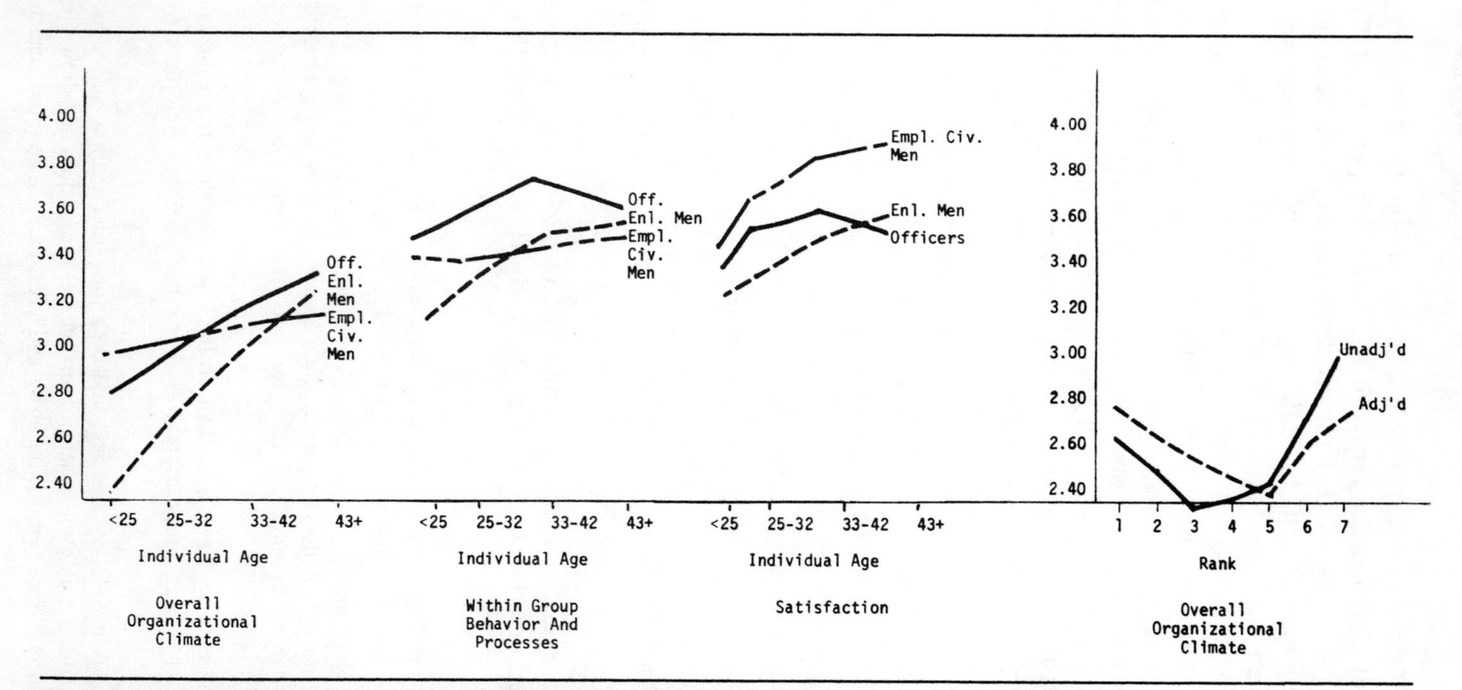

Figure 6: EXPERIENCED ORGANIZATIONAL PRACTICES BY AGE AND RANK

moving about than is the typical civilian) and that it provides them an insufficient opportunity to control their personal lives.

–Navy men, in far greater proportions than civilians, feel enmeshed in a large bureaucracy.

The picture changes somewhat as one moves from a consideration of the total Navy sample to a comparison of two of its major functional subunits, the fleet and the shore establishment. Figure 5, which contains total naval sample data, also presents line-graph profiles of the data from ship- and shore-based respondents. Using the 40 and 60 percentile points once more as demarcating a roughly "normal" range, we find distinct differences:

–While the shore establishment is, on all measures except Lower Level Influence, within the normal range, the fleet is, with two exceptions, below the 40th percentile on all measures.

–The differences between ship and shore are most pronounced on Motivational Conditions (an organizational climate measure), with ship respondents reporting levels worse than three-fourths of the national industrial array, whereas shore respondents fall near the median.

–On certain other measures, ships fall at low percentile points also, with somewhat smaller differences from shore only because the latter are themselves somewhat low:

–All other measures of organizational climate.

–The general satisfaction index.

THE EFFECTS OF AGE AND UNIT LEVEL

Careful scrutiny reveals that for Navy men (unlike civilians), preferences in the work setting rose or improved with age. These age effects seem even more pronounced for experienced practices than for preferences, and rank has effects independent of those associated with age. These findings suggest that the level of one's unit in the organizational hierarchy, rather than one's own rank, appears to be the more urgent consideration.

The data presented in Figure 6 indicate that there is, for organizational climate and within group behaviors and processes, a rise in quality of experience with age that (a) is steeper for enlisted Navy men than for officers and (b) scarcely exists for civilians. Satisfaction displays similarly steep rises with age for all three groups, however.

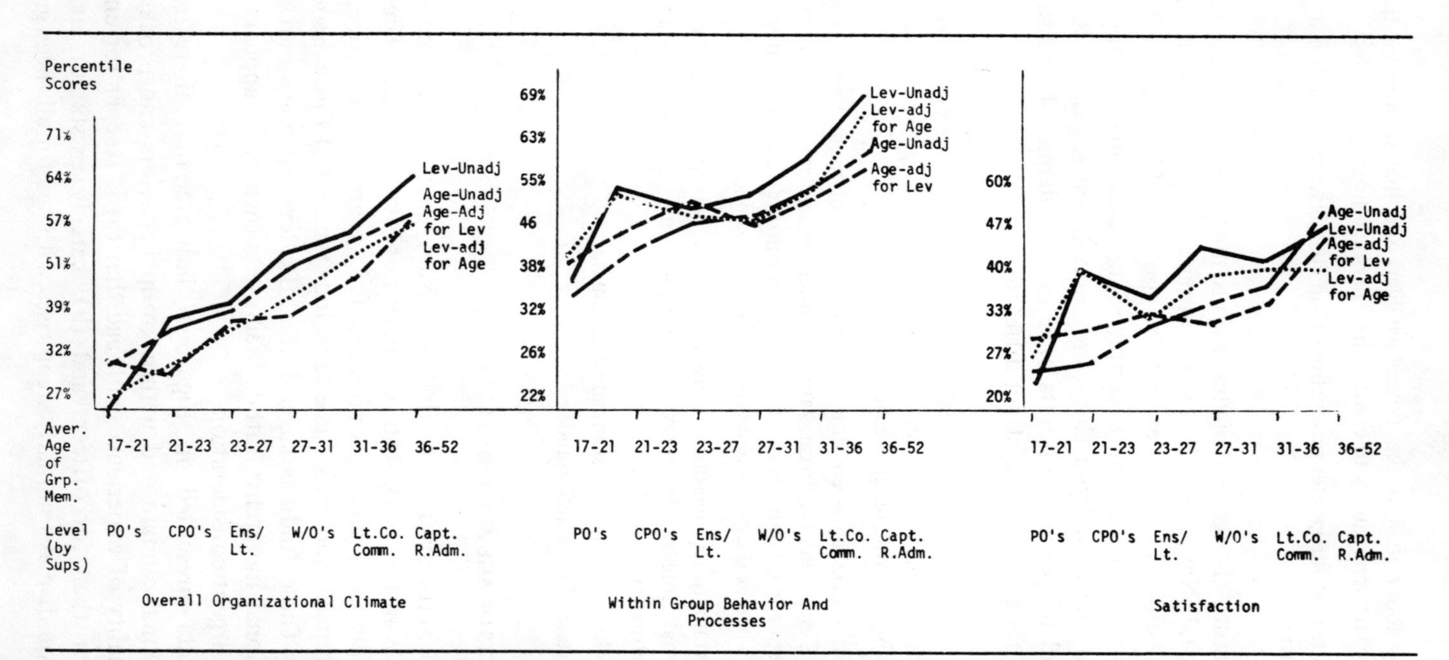

Figure 7: EXPERIENCED ORGANIZATIONAL PRACTICES BY AVERAGE AGE OF GROUP MEMBERS AND UNIT LEVEL

The earlier finding, that personal rank relates significantly to experienced practices independently of such considerations as age, is confirmed by the data in Figure 6. The interpretation offered as plausible—that part of the rise with age reflects a steady rise in positiveness with rank—is not confirmed, however. The present chart illustrates that the effect of rank, both raw and adjusted to remove the effects of education and self-selection as well as age, is curvilinear, first declining and then rising.

A technical report in the series from which these findings are drawn looked at some of these same effects from an organizational, rather than an individual, viewpoint (Franklin, 1973). In this situation, in which age or rank is associated with *organizational* differences, an individual may himself be young or lower in rank, yet a member of a group which is, on the average, older and headed by a person whose rank indicates that the unit which he heads is well up in the structure. The treatment which the young person receives in this latter situation may well be different from that received by a person of the same age in a younger, lower-status group.

Figure 7 presents data similar to those presented in relation to individual age and rank; however, *average* age of the group and *supervisor's* rank provide the basis for an analysis of *group means* upon clustered variables. Here we see that experienced practices for whole groups rise in positiveness with *average age* of group members in much the same way as obtained for individuals. Little change in these curves occurs when one adjusts for the effect of unit *level* (defined in terms of the supervisor's rank). Unlike individual rank, group level does seem to be associated with a relatively linear rise in the quality of experienced organizational practices, a rise which is only moderately reduced by controlling for average age.

These findings would appear to justify the conclusion that a Navy man's experience is at least in part a function of (a) his own age, (b) the average age or seniority of the persons in the group to which he belongs, and (c) his group's level or status in the organization. Combining these characteristics, one may surmise that an older person, in a group whose average age is similarly older and which is supervised by a person of higher rank, will experience by far the best organizational conditions. At the opposite extreme, the most unfavorable conditions will be experienced by young Navy men in lower-echelon groups whose members are, like themselves, young.

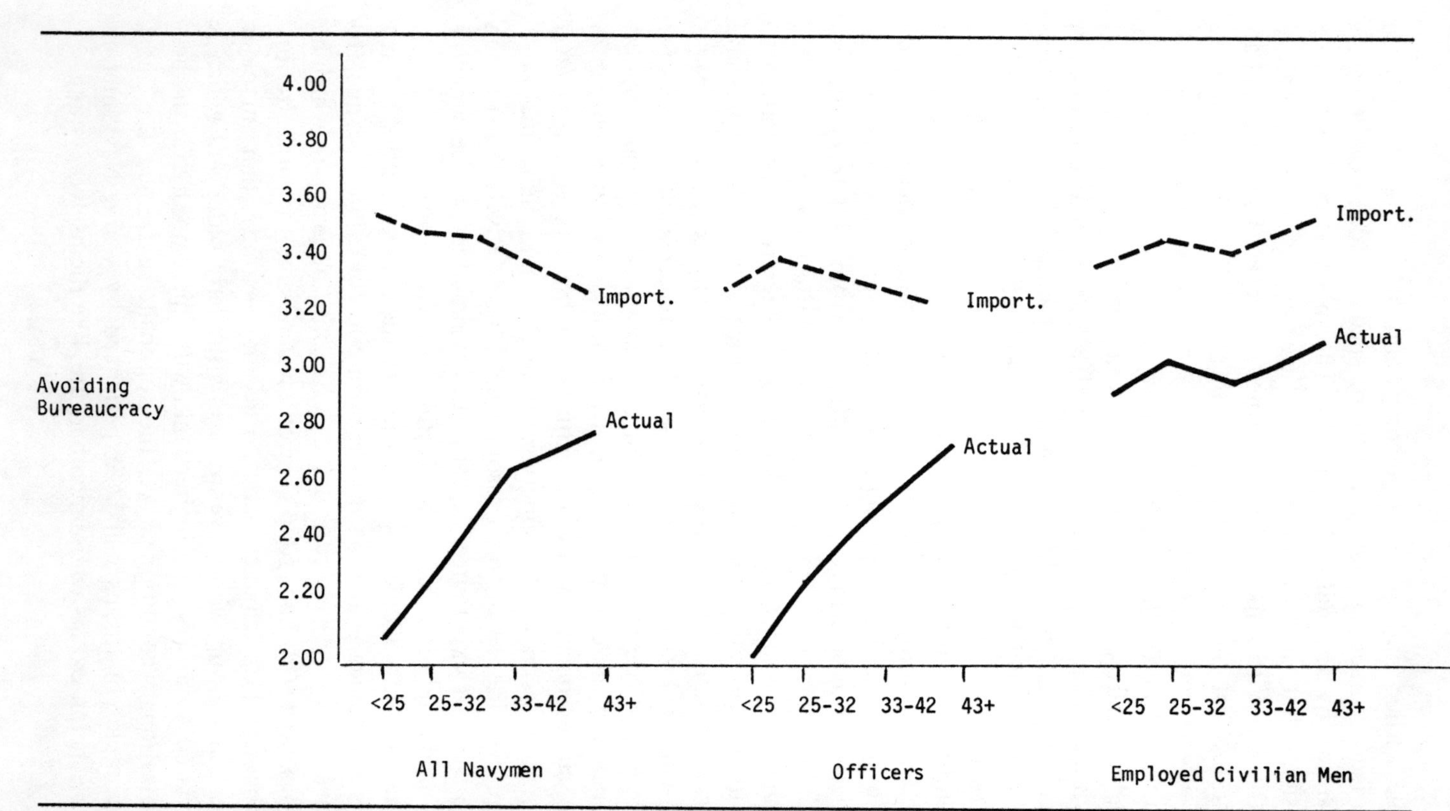

Figure 8: BUREAUCRATIC ENCOUNTERS: IMPORTANCE AND EXPERIENCE OF AVOIDING BUREAUCRACY BY AGE

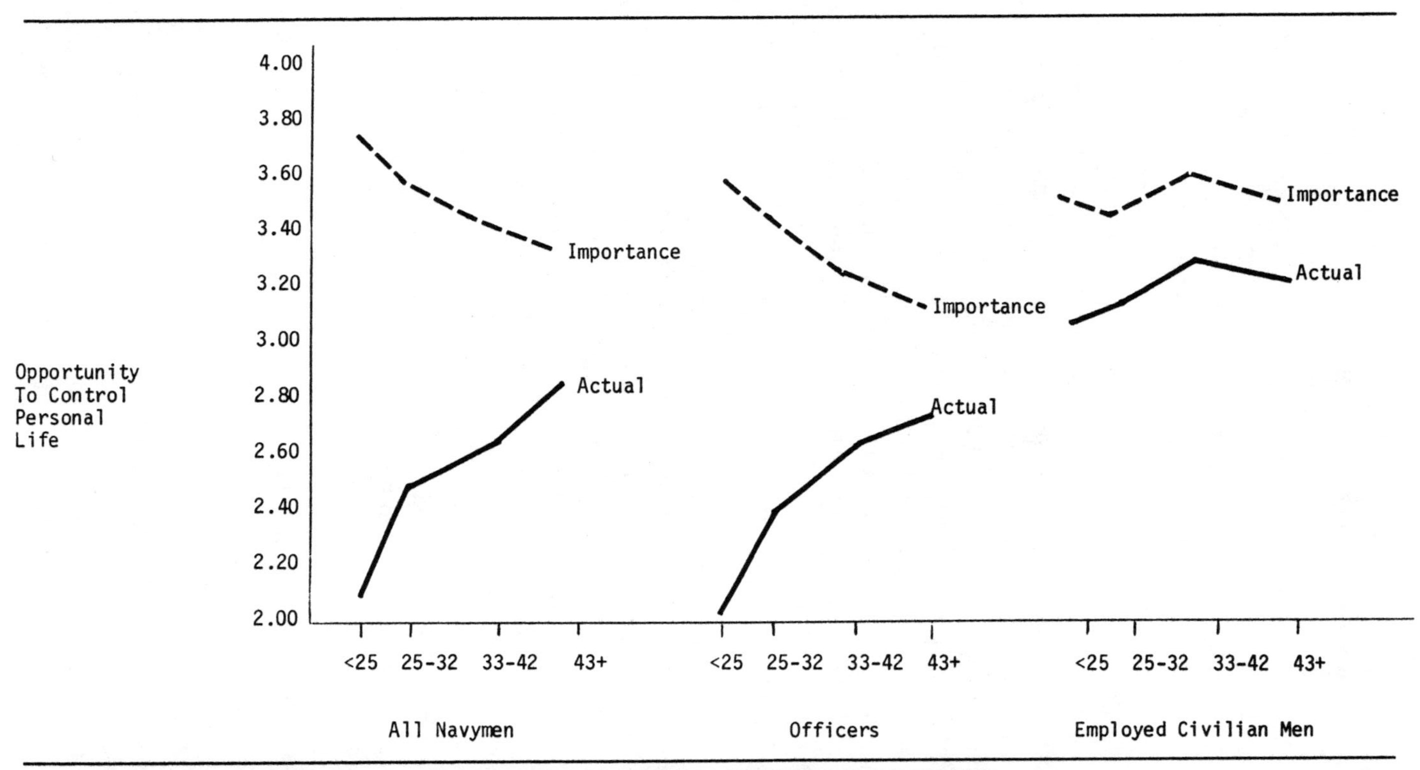

Figure 9: PERSONAL INDEPENDENCE: OPPORTUNITY TO CONTROL PERSONAL LIFE, BY AGE

PERSONAL INDEPENDENCE: BUREAUCRACY AND ONE'S PERSONAL LIFE

One issue stands out with such great importance that its relation to age has been isolated in this section for separate treatment. Stated most generally, it is personal freedom and independence, the ability to live the personal aspects of one's life reasonably free from external and bureaucratic constraints. Two measures were used in this study to tap the experience and importance of these characteristics: (a) a three-item index of the extent to which one is able to avoid endless referrals, red tape, and unexplainable rules (a high score therefore represents high independence) and (b) a single-item measure of opportunity to control one's personal life.

Both the actual experience and importance of these characteristics are presented in Figures 8 and 9 for all Navy men, Navy officers, and employed civilian men. The findings are clear and compelling: although Navy men and civilians attach approximately the same levels of importance to these qualities, only civilians experience what may be termed an acceptable or satisfactory degree of them. Young Navy men, furthermore, whether officer or enlisted, report an importance-experience gap of very large proportions.

Somewhat similar effects occur with respect to educational level. Actual experience and importance ratings for the Avoiding Bureaucracy index are presented in Figure 10 for enlisted Navy men, officers, and employed civilian men. Several facts are apparent from these data. First, for civilians the actual experience curve, like the importance curves, is flat and comparatively high, indicating that little difference in bureaucratic encounters is associated with educational level. Second, the Navy actual experience curves, for officers as well as enlisted men, are negatively sloped. In other words, despite more nearly common levels of aversion to bureaucracy, better educated Navy men report more frequent endless referrals, more occurrence of red tape, and a greater incidence of rules or regulations which no one seems able to explain than is reported by less well-educated persons.

Finally, the other "independence" measure–opportunity to control one's personal life–displays for officers a similar, rather strange, pattern (see Figure 11). The importance attached to being able to control one's personal life rises only slightly with education, a finding in no way surprising. Yet where most societies or social orders provide their technical-educational elites with more, not less, personal freedom, the reverse appears to be true among Navy officers. That the situation is decidedly different from the aspirations and experience of comparable groups in the civilian world is indicated by curves presented for employed civilian men.

Conclusions

WHAT THE DATA TELL US ABOUT VALUES AND PREFERENCES REGARDING THE WORK SETTING

Is there an organizational "generation gap"; that is, do young persons today value and prefer something different from what those more senior prefer?

For preferred characteristics of the job, the answer must decidedly be "no." Young persons appear to attach greatest importance to the rather traditional values of personal independence and material success, a preference which they share with all other civilian, and nearly all Navy, age groups. In this connection, it is worth noting that *serving one's country* ranks very low in importance. Different from all other groups, Navy and civilian, are Navy men 43 years of age and older (enlisted men as well as officers), for whom service to one's country is more important, personal independence less important.

The response must also be "no" concerning preferred leadership styles. Preferences in this area appear to track actual experience (at a somewhat higher level), an actual experience which is partly situational and fortuitous, partly a function of rank.

The answer is "yes," however, in terms of adherence to or acceptance of autocratic beliefs. This rises rather sharply with age, despite the fact that both experience with and preference for nonautocratic behaviors from others rises with age. The gap in adherence to autocratic beliefs is largest for young versus older enlisted men. Despite their similarities in other areas, it is nearly as large for *older officers* versus *older enlisted men,* the former looking very much like younger officers.

Is educational level related to preferences and expectations?

The answer must be "yes," in relation to *some* aspects of what people want from a job. Greater education is associated with reduced

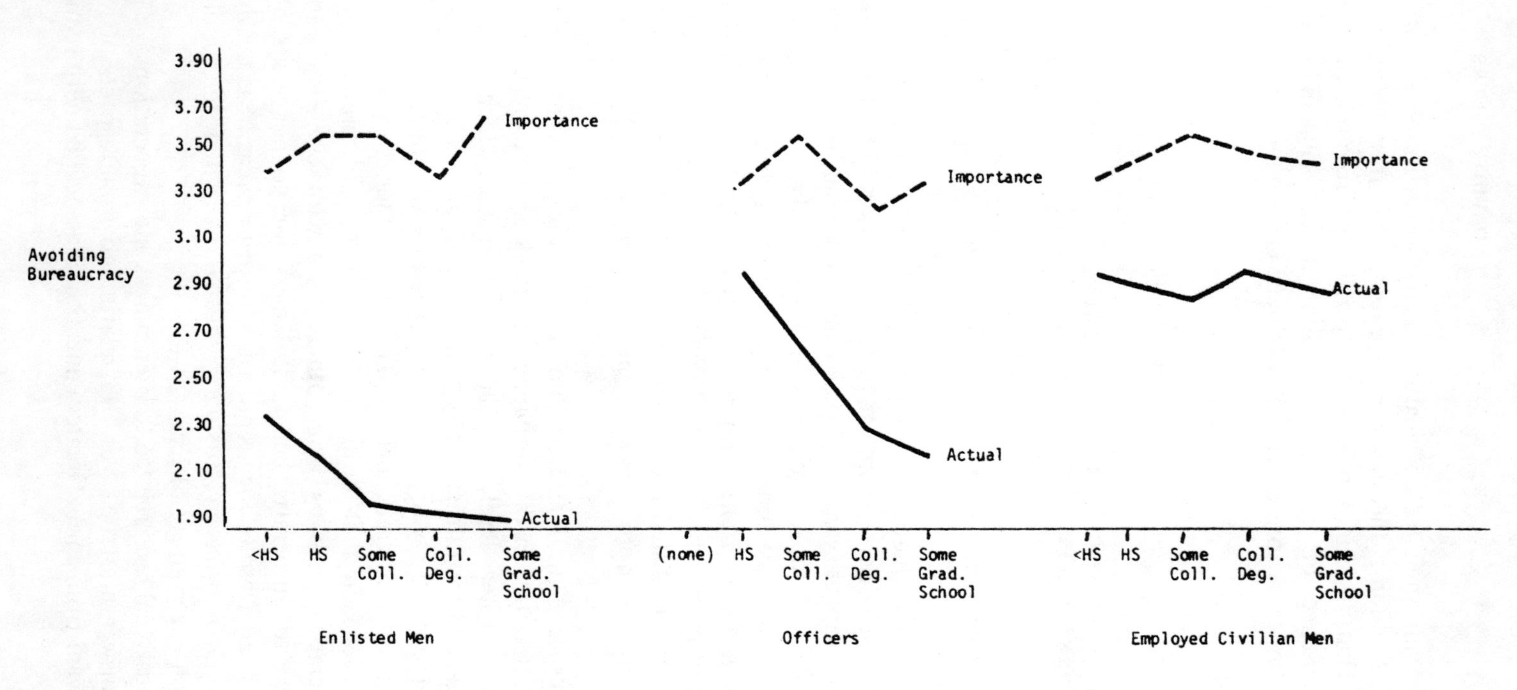

Figure 10: BUREAUCRATIC ENCOUNTERS: IMPORTANCE AND EXPERIENCE OF *AVOIDING* BUREAUCRACY, BY EDUCATIONAL LEVEL

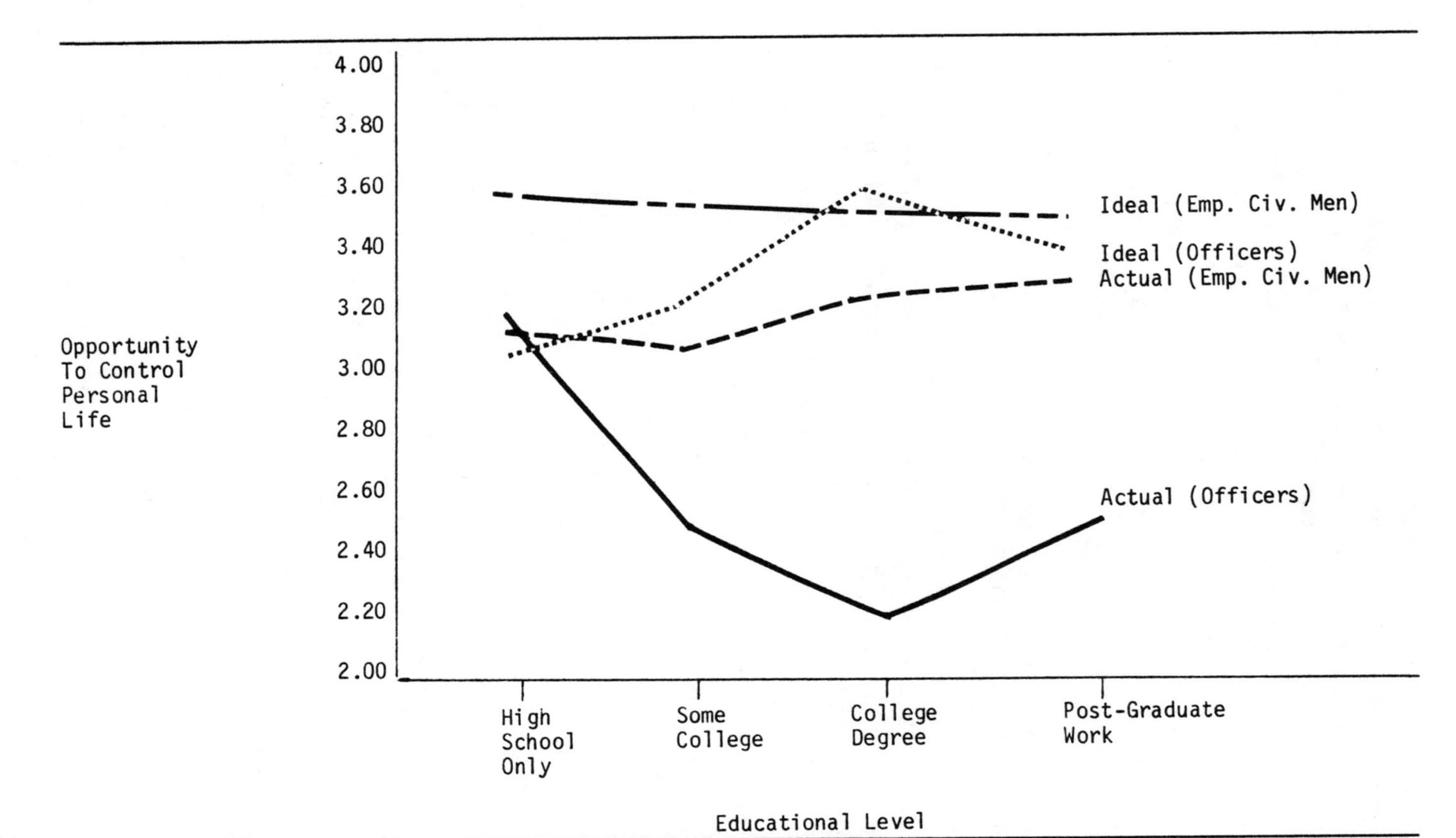

Figure 11: PERSONAL INDEPENDENCE: OPPORTUNITY TO CONTROL PERSONAL LIFE, BY EDUCATIONAL LEVEL

concern for economic issues, with less concern for serving one's country, and with enhanced concern about having challenging work. Among Navy men, it is also associated with the attachment of greater importance to personal independence.

The answer is also "yes" in terms of adherence to autocratic beliefs. This declines rather sharply with education, in this instance *paralleling* preferred and actual leadership practices.

The answer seems to be "no" in relation to preferred leadership practices. As with comparisons by age, rises with educational level appear largely to reflect the "tracking" of actual experience.

WHAT THE DATA SAY ABOUT THE NAVY AS A FUNCTIONING ORGANIZATION

1. The measures of organizational practices included in the survey represent a well-researched set of management principles. Appropriately structured, they form a picture or model of how an organization functions effectively. The data show that this model is reasonably valid for the Navy, as for civilian organizations, since:

(a) the various measures relate to each other as they should;

(b) the measures relate well to organizational criteria, especially retention rate.

2. Although the Navy as a whole falls approximately at the lower border of what is termed the "normal" range, this conceals a rather crucial difference. The shore-based units are well within that normal range; the fleet units are distinctly below it. The sole exceptions are the submarines, which resemble the shore units in quality or organizational functioning. Most of the more serious fleet problems appear to lie in organizational climate conditions and leadership behaviors, rather than in the intrinsic properties of jobs performed.

3. Much of the problem pattern occurs as well in, and perhaps ties critically to, a perceived undue absence of personal independence, in the form of bureaucracy and an unnecessary intrusion into Navy men's personal lives.

4. Like the organizational climate and leadership problems, this personal independence shortage is

(a) age-related—the favorability of practices experienced by Navy men rises with both personal age and the average age of the group to which one belongs.

(b) rank-related–for enlisted men, experienced conditions steadily decline from E1 to E5, then rise to a peak at E7.

(c) unit-level related–experienced conditions improve steadily with the rank of one's supervisor.

Note

1. The S.O.O. national array, rather than the civilian cross-section from the present study, is used for charting and percentile purposes because of the much larger number of cases contained in the former (more than 20,000). Analyses indicate that the civilian cross-section subsample of industrial employees (considered to be the best comparison base in the present instance from the overall cross-section) is not appreciably different from the S.O.O. national array. The mean index value of the two civilian comparison bases is different by only .07 of one scale point, and the profile of indices intercorrelated (rank order coefficient) .93.

IV.

THE MILITARY FAMILY

TRENDS IN FAMILY PATTERNS OF U.S. MILITARY PERSONNEL DURING THE 20th CENTURY

Nancy L. Goldman

Over the last century, the U.S. military has undergone a demographic transformation. The armed forces have gradually shifted from an institution in which the majority of the personnel were single to one in which a majority are married.[1] Moreover, the patterns of marriage and the age structure of the military are such that military families while on active duty have children at home. Thus, there has been a pronounced shift in the internal structure of the armed forces as they emerge as a familistic institution.[2]

Basic trends reflect the developments in the larger society: the percentage of persons living in family units has increased; the growth of medical technology has reduced childless marriages and has made it possible to limit the size of one's family as well. Second, there are certain trends in the military establishment. There is a lower divorce rate in the military, and those who do divorce tend to be selected out. The average age of active-duty personnel has been reduced through earlier retirements. Third, basic trends reflect the fundamental transformation of the position of the military in an advanced industrial society in which military personnel who are career or quasi-career personnel demand and are given the same style of life as civilians. Trends toward the all-volunteer force increase this pressure for all but the cadres who will serve one term or limited lengths of time.

Historical Background

Because of military regulations and practices, the majority of active-duty personnel at the turn of the century were unmarried; those who were married were predominantly older officers. Until World War II, "military wife" was synonomous with "officer's wife," since few enlisted men were married, and, if married, their families were of little concern to the institution. However, it was considered a necessity for an officer above the age of 30 to have a wife, since she filled an important role in his social life (Janowitz, 1960, 1971b; Karsten, 1972). Although officers in the early 1900s typically were in their thirties at the time of marriage, some married shortly after graduation from the academies, and the military wedding was a special event. Often either or both the bride and groom were children of officers. In her letters, Frances Roe (1909), wife of an army officer in the late 1800s, described a frontier post wedding of 1887:

> It was so strictly a military wedding—at a frontier post where everything is of necessity "army blue"—the bride a child of the regiment, her father an officer in the regiment many years, and the groom a recent West Point graduate, a Lieutenant in the regiment. . . .
>
> At the rear end of the hall stood two companies of enlisted men . . . in shining full-dress uniforms, helmets in hand. The bride's father is Captain of one of those companies and the groom's a lieutenant in the other. . . . One became impressed with the military completeness of the whole scene.

Progressively over the 20th century the proportions of both officers and enlisted men who were married increased, and officers, in particular, married at an earlier age.[3] Enlisted men, originally required to be unmarried at the time of their initial enlistment, most often remained so during their term of duty. Those who married were mostly older men who had made a career commitment to military service.

Men enlisting in the Army for the first time were required to state in writing that they were unmarried and had no children. Age requirements during the late 19th century stipulated 16-36 years of age for such personnel, and a 5-year term of service. Many enlisted men, however, deserted and never fulfilled the 5-year obligation, a frequent reason being the decision to marry. As early as 1847, Army Regulations stated:

> No man having a wife or child shall be enlisted in time of peace without special authority from the General Headquarters through the Superintendent. This rule is not to apply to soldiers who "re-enlist." [U.S. War Department, 1947:135]

Such regulations were maintained for the post-Civil War enlisted soldier. By 1892 Congress favored the unmarried law which provided that any unmarried enlisted man under 30 years of age could present himself to the Board of Examiners for promotion, thereby opening the door for the unmarried man to obtain a commission. Despite this opportunity having been extended to the single enlisted man, few are reported to have benefited from it.[4]

Therefore, between restrictive Army regulations about the marital status of enlisted recruits and poor housing conditions, there were few married men on the frontier posts—young or old.[5] Furthermore, the Army of the 1890s was often referred to as "an army of deserters," and those who enlisted were held in disrepute by the public, making it unlikely that a self-respecting "family man" would subject a wife and children to such a way of life. General Crook described this situation in the *Army Navy Journal* (October 11, 1890):

> The moment a citizen dons the uniform of a private, however, no matter what his previous social position may have been, he is instantly ostracized by the public. . . . A man loses his pride and self-respect when he finds he is despised by the people he meets, that he is shunned by former associates and is no longer regarded as their social equal.

Enlisted men who served repeated terms of duty—those with career intentions—were more likely to marry than those volunteering for one term only, who often used the Army as a temporary means of employment during a period when other sources of employment were scarce, as in 1893 during the labor uprisings (Ganoe, 1924). In the late 19th century, the majority of regulars left the Army, if not during their first enlistment term, at the end of it. The Act of June 16, 1890, had offered an honorable discharge from service at the end of 3 years (out of a 5-year term required in 1890) or by purchase before that time, effectively shortening the term of service to 1 or 2 years. Consequently, large numbers of men, including noncommissioned officers, left the service. An officer, in a letter to the *Army Navy Journal* (October 11, 1890) stated:

> It is a deplorable fact that a large number of our best noncommissioned officers are daily taking advantage of the late General Orders to be discharged.

In 1893, a Senate bill provided that in peacetime privates over 30 years of age who had served for 10 years or more could not reenlist.

Thus, with the possibility of being dismissed after 10 years of service and hardship, many did not reenlist—particularly those who had married while in service. Frequent excuses were "my wife wanted a permanent home of her own and children." Many left rather than bring a wife to live in such isolation, and only a very few enlisted men, mostly sergeants, remained in the Army for more than one enlistment period in hopes of receiving a commission, since such an event happened rarely, although the guidelines set forth in Army General Orders indicated means by which regulars could apply for a commission (U.S. Congress, 1892). There was little sentiment about the Army as a lifelong profession for an enlisted man, especially since it was ruled that men who remained privates for 10 years were barred from reenlistment.

For the officer's family, life was also difficult, despite the fact that by the turn of the century the Army posts were less physically isolated as a result of the railroad and the automobile. These allowed the Army to become centralized in regimental units rather than remaining spread out in different frontier posts. These frontier posts barely provided for wives and children of young officers. Life in the U.S. Army in the 1870s was described by Martha Summerhayes (1911:71) in her memoirs as one of "glittering misery." In answer to her complaint about the inadequate, cramped housing allotted a lieutenant, her husband responded, "Why, Martha, did you not know that women are not reckoned on at all at the War Department?" (p. 16). During this period, according to Army regulations, a lieutenant's housing allowance provided one room and a kitchen; each rank allotted another room; a colonel had a fairly nice house (p. 16). It was important that the senior officer of the post be a married man, and by 1890 it was less uncommon for officers to marry. Of the mean strength of 24,726 officers and enlisted personnel, 18 officers and 34 enlisted men were reported to have married that year; and the following year, out of a mean strength of 25,008 men, 11 officers and 31 enlisted men married. There were 87 children born to wives of officers and 349 to those of enlisted personnel (U.S. Congress, 1889-1890; 1890-1891). Nevertheless, by 1907 there were still few school-age children on the posts; few posts provided schools. At 18 stations of U.S. troops, there were no school-age children reported, and the question of schools had not even been considered. Only one post reported as many as 40 children, and most had fewer. The majority of these were children of enlisted men and civilian employees. Of 101 posts reported in "Educating Children at Military Posts," only 15 post schools are mentioned (U.S. Congress, 1908).

Although there were no legal restrictions against the marriage of officers, they were discouraged from marrying before they were 30 to 35 years old. Karsten (1972) reports that, in the Annapolis class of 1871, of 38 graduates who were on active duty in the early 1900s, at the time of the study, 31 had married 10 to 15 years after graduation. Biographies of military figures also reveal the more mature age at which officers married during this period in contrast to the post-World War II period, when many men married immediately after graduation from the military academies, marriage before graduation being strictly forbidden; those who did marry before graduation were forced to leave.[6] As late as 1963 this rule was under discussion (*Navy-Air Force Journal,* October 12).

The history of dependency benefits summarizes the gradual development of welfare benefits accorded families of servicemen until World War II and the growing concern of the military for the families of their personnel. During each war, allotments for dependents of men in service were provided, but it was not until the Spanish-American War that the allotment act passed by the 55th Congress to meet war emergencies was extended to the postwar period, becoming a permanent Army service in peacetime as well. These privileges, nevertheless, were not extended to *officers'* families until World War I, at which time the first system of family allotments established by the 65th Congress as a war measure accorded rights to commissioned officers as well as to enlisted men. Not until 1942, with the passage of Public Law 490 by the 77th Congress, were dependency benefits provided, in addition to the allotment of the serviceman's pay. These benefits included monetary provisions for dependents and some obstetric care.

Trends Since World War II

FAMILY STATUS AND AGE PATTERNS IN THE U.S. MILITARY

Following World War II, the U.S. Armed Forces experienced a steady rise in the percentage of total active-duty force who were married and had children; from 1953 to 1974 there was a 48%

Table 1. NUMBER AND PERCENTAGE OF ACTIVE-DUTY MALE PERSONNEL WHO ARE MARRIED

	Number of Males on Active Duty	1953-1974 Number of Males Married	Percentage Married
1953	3,343,900	1,279,295	38.3
1955	2,891,529	1,207,156	41.7
1956	2,762,696	1,214,750	44.0
1957	2,763,483	1,280,776	46.3
1960	2,447,040	1,276,308	52.2
1965	2,678,878	1,285,799	48.0
1968	3,436,769	1,588,942	46.2
1969	3,394,767	1,566,589	46.1
1970	2,927,881	1,506,966	51.5
1971	2,569,586	1,387,326	54.0
1972	2,295,776	1,256,930	54.8
1973	2,158,477	1,217,432	56.4
1974	2,104,969	1,197,650	56.9

SOURCE: derived from tables of progress reports and statistics of the Office of the Secretary of Defense.

increase in the percentage of all uniformed personnel who were married (see Table 1). In 1953, 38.3% were married; and this figure rose to 52.2% in 1960. It dropped to 46.2% in 1968 as a result of the Vietnam buildup; but the rising trend began again in 1970 and by 1974 reached 56.9%. This rate of growth between 1953 and 1974 exceeds that of civilian society, which between 1950 and 1972 increased its proportion of married males who were 18 years of age and older by only 10%, as compared with the 38% increase in the percentage of male personnel in the military who are married (U.S. Bureau of the Census, 1974).[7] This difference in growth rates emphasizes the trend of the military toward becoming a familistic organization with the proportion of its members who are married rapidly approaching that of those who are 18 years of age and older in the larger society (see below).

The number of dependents per active-duty male have increased with the increase in the percentage of personnel who are married. (See Table 2.) In 1953 there were 0.85 dependents per uniformed male. The number of these dependents rose to the high of 1.55 in 1964, remaining at this level until 1968 and 1969, when it decreased to 1.3 as a result of the mobilization for the Vietnam War of many young, unmarried enlisted men. (Table 3 shows that during these 2 years the median age of active-duty personnel was 22.6, the lowest age in the entire 1953-1974 period.) However, by 1974 the number

Table 2. NUMBER AND PERCENTAGE OF DEPENDENTS OF ACTIVE-DUTY PERSONNEL 1953-1973 (Wives, Children, Others)

Year	Total Number (millions)	Number per Military Person	Number per Military Family
1953	2.89	.85	2.26
1954	2.59	.82	2.23
1955	2.89	.99	2.39
1956	3.01	1.08	2.48
1957	3.24	1.16	2.53
1958	3.43	1.33	2.75
1959	3.54	1.43	2.81
1960	3.73	1.50	2.92
1964	4.01	1.55	2.99
1965	3.81	1.55	2.89
1968	4.34	1.3	2.87
1969	4.36	1.3	2.87
1971	3.95	1.50	2.82
1972	3.40	1.53	2.79
1973	3.33	1.54	2.74
1974	3.28	1.55	2.74

SOURCE: derived from "Active Duty Military Personnel and Their Dependents," Tables of Directorate for Statistical Services, Office of the Secretary of Defense.

of dependents per active-duty member returned to its highest level to date of 1.55.

In 1974 the military family, which averaged 2.74 dependents, was smaller than in other years after 1958, and especially smaller than in 1964, when there were almost 3 dependents per family (2.99), indicating a trend toward smaller families. It was found in *Army Families* (Bennett et al., 1974) that between 1966 and 1969 there was a 47% increase in the number of army officers and a 50% increase in the number of enlisted men who had one or no children, while the number of families having two or more children decreased. For society at large, the percentage of families having no children

Table 3. MEDIAN AGE OF ACTIVE-DUTY MALE PERSONNEL 1960-1973

1960	24.2
1964	24.3
1965	24.0
1968	22.6
1969	22.7
1971	23.0
1972	23.1
1973	23.9

SOURCE: *Military Market Facts Book, 1974* (U.S. Army Times Publishing Co., 1974).

Table 4. PERCENTAGE MARRIED OF OFFICERS AND ENLISTED MEN BY SERVICE

	Department of Defense			Army			Navy			Air Force			Marine Corps		
	Total	Off.	Enl.	Total	Off.	Enl.	Total	Off.	Enl.	Total	Off.	Enl.	Total	Off.	Enl.
1953	38.3	81.1	33.3	35.2	80.9	30.7	37.2	78.3	32.5	46.4	84.4	40.5	30.4	72.9	26.7
1955	41.7	78.0	36.9	39.4	81.1	34.4	40.7	73.3	36.7	47.8	78.6	42.7	29.7	71.8	25.4
1957	46.3	82.9	41.4	46.5	88.1	41.5	40.3	74.8	36.2	54.0	83.9	48.6	31.4	76.9	27.2
1960	52.2	84.9	47.5	49.9	87.2	45.2	42.8	76.6	38.7	65.0	88.1	60.6	37.3	78.2	33.2
1961	49.5	82.3	45.0	44.7	79.4	40.4	42.4	76.1	38.4	62.9	88.2	58.3	37.0	79.7	33.0
1965	48.0	80.7	43.5	41.4	81.8	36.4	42.0	74.0	38.0	63.9	84.5	60.1	35.6	74.5	31.9
1967	42.9	76.9	38.5	40.8	74.6	36.9	41.1	75.7	37.0	54.0	82.0	49.1	24.2	66.8	20.6
1970	51.5	80.8	47.0	50.2	80.0	45.9	44.5	78.0	40.3	67.7	87.7	63.7	26.9	59.6	23.4
1971	54.0	81.9	49.5	55.5	84.3	50.9	46.9	79.6	42.5	63.0	82.8	59.1	34.8	69.1	30.9
1972	54.8	81.9	50.4	54.4	82.3	49.8	48.2	80.9	43.8	65.9	84.0	62.3	36.4	70.0	32.8
1973	56.4	83.5	52.1	55.6	85.5	50.8	50.4	80.8	46.2	67.3	84.7	63.8	39.7	75.1	35.9
1974	56.9	83.1	52.7	55.4	83.5	50.9	52.0	81.2	47.9	67.7	84.9	64.2	40.3	76.3	36.4

SOURCE: derived from tables from the Department of Defense, Directorate for Information Operations, December 31, 1953-March 31, 1974.

increased slightly, from 44.7% in 1960 to 44.2% in 1969, reaching 45.6% in 1973, while those with one child decreased from 18.5% in 1960 to 18.1% in 1969 and subsequently increased to 19.3% in 1973. Families consisting of three children decreased from 11.1% in 1960 to 10.5% in 1969 and dropped to 9.7% in 1973.

FAMILY PATTERNS OF OFFICERS AND ENLISTED MEN IN THE FOUR SERVICES

The trend in the percentage of married personnel in the armed forces since World War II has risen for both officers and enlisted men in each of the four services; however, the change has been much greater for enlisted men than for officers (U.S. Department of Defense, 1974). From 1953 to 1974, the percentage of officers of the total armed forces who were married rose from 81.1% to 83.1%, while for enlisted men the change was from 33.3% in 1953 to 52.7% in 1974. (See Table 4.)

When the four services are compared for the total of married personnel, the Air Force in 1953 as well as 1974 had the highest percentage. In 1953 the Air Force had 46.4% married, and in 1974 there were 67.7% married. The Marine Corps, on the other hand, has had the smallest concentration of married personnel of all of the services: 30.4% in 1953 and 40.3% in 1974. It has been the only service in which, as of 1974, the unmarried continued to outnumber the married. During this period, the Army has undergone the greatest transformation, from a service having a total of 35.2% married personnel to one with 55.4% married in 1974–a 57% increase. (The Air Force increased by 46%, the Navy by 40%, and the Marines by 33%.)

When the services are compared for percentage of enlisted personnel who were married, it is found that, of the four services, the Army also had the greatest increase. In 1953, 30.7% of its enlisted men were married, and in 1974 there were 50.9% married, a 66% increase. The Marine Corps's enlisted grades increased the least: from 26.7% married to 36.4%–a 36% increase, compared to the 50% increase of the Air Force and the 40% of the Navy.

When, however, the services are compared for the percentage of officers who were married, the Marine Corps is found to have had the greatest increase, since changes for the other services were slight. In 1953 there were 72.9% Marine Corps officers married; in 1974, 76.3%.

Table 5. NUMBER OF CHILDREN PER MILITARY FAMILY BY SERVICE FOR SELECTED YEARS 1953-1974

	Department of Defense			Army			Navy			Marines			Air Force		
	Off.	Enl.	Total	Off.	Enl.	Total	Off.	Enl.	Total	Off.	Enl.	Total	Off.	Enl.	Total
1953	1.50	.88	1.02	1.48	.81	.94	1.42	.92	.10	1.41	.75	.88	1.59	.99	1.14
1954	1.48	1.02	1.12	1.44	.90	1.01	1.51	1.16	1.22	1.58	1.11	1.22	1.48	1.07	1.18
1955	1.61	1.19	1.28	1.63	1.13	1.24	1.51	1.15	1.22	1.48	1.08	1.17	1.65	1.29	1.37
1957	1.73	1.31	1.40	1.68	1.22	1.31	1.72	1.39	1.45	1.61	1.23	1.31	1.78	1.37	1.46
1959	1.95	1.59	1.66	2.00	1.59	1.67	1.83	1.54	1.59	1.83	1.44	1.51	1.99	1.63	1.71
1960	2.07	1.71	1.78	2.19	1.86	1.92	1.93	1.57	1.64	1.93	1.54	1.61	2.04	1.68	1.76
1965	2.03	1.80	1.85	1.96	1.57	1.65	2.01	1.69	1.75	2.06	1.67	1.74	2.10	2.06	2.07
1970	1.75	1.49	1.55	1.63	1.30	1.37	1.79	1.46	1.52	2.04	1.29	1.45	1.81	1.79	1.79
1971	1.75	1.56	1.59	1.71	1.46	1.51	1.80	1.48	1.54	2.17	1.52	1.65	1.70	1.74	1.73
1973	1.81	1.57	1.62	1.86	1.53	1.60	1.80	1.56	1.61	2.00	1.37	1.48	1.74	1.64	1.62
1974	1.75	1.57	1.61	1.70	1.55	1.58	1.81	1.53	1.59	1.98	1.35	1.47	1.75	1.66	1.67

SOURCE: derived from statistical reports of the Office of the Secretary of Defense.

Although a comparison by pay grade of the percentage of officers and enlisted men who are married for all four services has not been made, the study of Army marriage trends shows that, with few exceptions for the years 1966, 1969, and 1973, the percentage of both Army officers and enlisted men who were married was higher for higher pay grades (Bennett et al., 1974). In addition, the officers in the three top grades had a slightly higher proportion who were married than those in the four top enlisted grades. However, the percentage married of the lower enlisted grades was much smaller than that of the lower officer ranks, since members of these lower enlisted grades are usually younger and have not made the career commitment that officers have made. While over 50% of the 2nd lieutenants were married, approximately only 18% of the E1 and E2 grades were married.[8]

Family patterns of officers and enlisted men in all the services show an increase in the average number of children per military family. (See Table 5.) However, in 1974 there were slightly fewer children per military family than in 1973, indicating a recent trend toward smaller families. Officers have consistently had more children than enlisted men. In 1953 officers had an average of 1.5 children per family, compared with .88 child per enlisted man's family. The figure reached its high point for both sectors in 1960, when officers had an average of 2.07 children per family and enlisted men had 1.71. Since then, the number of children has been decreasing; in 1974 officers had 1.75 children per family to the 1.57 of enlisted men. Thus the difference in number of children for the two groups has virtually disappeared, the enlisted men's having undergone the greater transformation in the size of the nuclear family.

When the number of children per family is examined by service for the period 1953-1974, the data show that the officers and enlisted men of the Air Force consistently had more children per family than the other services. Those of the Marines had the fewest. In 1974, however, the officers of the Marine Corps had the most children, Navy officers came next, followed by the Air Force. Army officers had the least. Among the enlisted ranks, the Air Force had the largest number of children, the Army the next largest, and the Marines the fewest.

When examined by pay grade for 1973, the families of middle-ranking officers of all services (grades 4-6) had the largest families, with approximately 3 children per family (U.S. Army Times

Publishing Co., 1974). Conversely, the junior officers and the enlisted men in grades E1-E4 had an average of less than one child per family.

FAMILY PATTERNS IN THE MILITARY AND THE GENERAL POPULATION

As already discussed, the marriage trends for the active-duty military community reflect those of the larger society to the extent that the marriage rates have increased between 1950-1953 and 1972 for both groups (see Table 6). However, the rate of increase for the military during this period has exceeded that for the civilian. The larger society increased from 68% married in 1950 to 74.8% in 1972, while the armed forces increased from 38.3% married in 1953 to 54.8% in 1972–a 10% increase for the former group and a 43% increase for the latter.

The officer segment of the armed forces has consistently had a somewhat higher percentage of married–81.1% in 1953 and 81.9% in 1972–than the general population. Marriage has always been considered vital to the career advancement of officers, while divorce has been extremely detrimental. Since World War II, social services have been provided by the military to prevent marital disruptions. Those who did divorce usually left the military. Consequently, divorce rates here have been much lower than in the civilian sector. Williams (1974) found that in 1970 only 1.2% Air Force officers were divorced, as compared with 3% for the larger society.

Marriage rates for enlisted men in the armed forces, on the other hand, despite their increase since 1953, continue to remain well below those for civilians. While in 1953, 33.3% of enlisted men were

Table 6. PERCENTAGE MARRIED–DEPARTMENT OF DEFENSE AND GENERAL POPULATION

	Department of Defense			
	Total	Officers	Enlisted Men	General Population
1950-1953*	38.3	81.1	33.3	68
1960	52.2	84.9	47.5	69.1
1970	51.5	80.8	47.0	75
1972	54.8	81.9	50.4	74.8

*1950–year of civilian data; 1953–year of Department of Defense data.

SOURCES: Department of Defense and *Statistical Abstract, 1974* (Washington, D.C.: U.S. Government Printing Office, 1974).

Table 7. 1973 MEDIAN AGE DISTRIBUTION: GENERAL AND MILITARY POPULATION

	Children	Wives	Males (18 and older)
General population	10.5	40.7	40.0
Military	5.3	23.0	23.9

SOURCES: *Statistical Abstract, 1974* (Washington, D.C.: U.S. Government Printing Office, 1974) and *Military Market Facts Book, 1974* (U.S. Army Times Publishing Co., 1974).

married, 68% of the general population were married; and in 1972, it was 50.4% for enlisted men, as compared to 74.8% for civilians. Thus, in 1972 there remained a 48% difference between the two groups, while in 1950-1953 this difference was 100%.

In terms of family size, the military not only has been catching up with the married in the general population but has had on the average more children under 18 years of age per family–1.62 children–than 11 other civilian occupational groups (U.S. Army Times Publishing Co., 1974). Yet the 3.68 members in the military family were fewer than in prior years; in 1970, the military family consisted of 3.87 persons. The size of the civilian family decreased more dramatically than the military during this period: from 3.14 persons per family to 3.01 (U.S. Bureau of the Census, 1974). Thus the military family, while smaller in 1973 than in previous years, has decreased proportionately less than the civilian family.

The military also differs in that the active-duty personnel and their dependents are younger than their equivalents in the general population. (See Table 7.) The children of military families are as a group much younger than those of the larger society; in fact, in 1973 their median age was half that of the civilian's, or 5.3 years compared with 10.5 years. Almost half (49.3%) of the children of the active-duty personnel were under 5 years of age, with 41.3% between 5 and 13.[9] The median age of the military wives was also much younger than that of the civilian wives–23 years compared with 40.7 years. Over 58% of the former and 12% of the latter were between the ages of 18 and 24. And fewer military wives than civilian wives were employed. For the civilian society in 1973, 42.2% of all women with husbands were present in the work force, while, in 1971, 22.1% of the Army officers' wives worked and 39% of the Army enlisted men's wives worked (U.S. Bureau of the Census, 1974; Bennett et al., 1974). The percentage of military wives who are employed outside the home has been increasing.

Conclusion

In the course of the 20th century, the U.S. military has been transformed from an organization composed of a high proportion of active-duty personnel who were unmarried to one in which the majority are married and have children. The most dramatic change in marital status since World War II has occurred in the enlisted category, although there has been an increase in the officer corps as well. Since 1960 the number of dependents of all military personnel have outnumbered the personnel. There has been a reduction in the median age of the active-duty personnel, since the retirement age has been lowered; consequently most members still have dependent children at the time of their retirement.[10]

In the 1970s, with the shift toward an all-volunteer service, the military is faced with the intensified lack of personnel recruitment. Since personnel are being recruited into a highly familistic environment, the armed forces have come to offer benefits which allow them to compete with the civilian sector for manpower and make them particularly attractive to men with families. The benefits for dependents of active-duty personnel and the early retirement age are especially notable. This situation is in marked contrast to that of the frontier days, or even those of World War I, when few if any provisions were made for dependents. The military is dealing not only with families but also with relatively young families. These operatives of family composition and the young age distribution supply one important dimension of the context of the new trends in the military profession.

Notes

1. Throughout this paper, "military personnel" refers to active-duty males.

2. This paper draws on a variety of statistical sources and has been able to make use of the extensive study of the demography of Army families, "Demographic and Social Characteristics" in *Army Families* (Bennett et al., 1974).

3. An examination of records at the U.S. Military Academy reveals that the percentage of their graduates who married at the Academy during the June of their graduation increased from 2% in 1900 to 13% in 1970. While these percentages do not take into account those who married outside the Academy and may reflect trends in marrying at the Academy, some indication of the change in marriage rates among the graduating class is provided. (At the request of the author, Judy Scott obtained the numbers of marriages and of graduates from the following sources: 1900-1974 records of the Protestant Chaplain's office, USMA; 1927-1974 records of the Catholic Chaplain's office, USMA; 1900-1927

records of the Sacred Heart Rectory, Highland Falls, N.Y.; USMA Register of Graduates, 1973; the Cadet Roster.)

4. By 1917, Army regulations had become more flexible about the marital status of enlisted men, specifying limitations for the line and "discouraging" their marriage rather than forbidding it (U.S. War Office, 1917:176): "The enlistment or reenlistment of married men for the line of the Army is to be discouraged and will be permitted only for some good reason in the public interest, the efficiency of the service to be the first consideration." As late as 1942 a general statement of conditions under which a man may enlist in the U.S. Army specified that he be "unmarried" (Ford, 1941).

5. A House executive document (U.S. Congress, 1890-1891) reports on the terrible housing and poor sanitary conditions which would make it most undesirable for family living.

6. See n. 3 above.

7. "Civilian" or "general population" refers to males 18 years of age and older in society at large.

8. It is interesting to note that in 1936 the percentage of enlisted men who were married increased by grade groups: 9.8% of the privates were married, 38.5% of the corporals, 66.5% of the sergeants, 80.7% of the first 3 grades. (Of all of the enlisted men, 21.6% were married.)

9. These data are adapted from the Group Research Project, *Army Families* (Bennett et al., 1974), which presents a comparison for these years for the Army and civilian society, and from a Department of Defense table entitled, "Active Duty Military Personnel and their Dependents 1972." The figures for 1950 are for the general population and those for 1953 are for the military population.

10. By 1973 the average age for nondisability retirements of both enlisted men and officers combined was 42.6 years, with an average of 22.6 years of service; the median age for officers was 55.9 years, with a median rank of major/lieutenant commander and an average of 22.2 years of service. For the enlisted men with nondisability retirements, the median age for retirement was 47.6 years, with a median rank of sergeant 1st class and an average of 22.0 years of service (Army Times Publishing Company, 1974). Before 1950 most careerists served 40 to 50 years. Since then, most have retired after 20 years of service. See also unpublished Department of Defense tables entitled "Retirements from Active Duty by Years of Service, Personnel Category and Type of Retirement," OASD (M&RA).

THE MILITARY FAMILY: ITS FUTURE IN THE ALL-VOLUNTEER CONTEXT

M. Duncan Stanton

Considering its importance as an institution, the family has not received its due on the national political front. As Vincent (1970) points out, the family more than any other major social institution facilitates social change by adapting its activities and structure to the changing needs of other social institutions and to society at large. Nonetheless, the family has usually been taken for granted, and efforts to gain recognition and support for it in policy decisions have normally been treated with indifference (Rue, 1973).

For years, policy toward and research on the family within the military were virtually nonexistent (Janowitz, 1960; Little, 1971). However, as the proportion of married to single servicemen began to increase during the cold war, the military began to revise its thinking on the subject. In recent years three-fourths of the officers, 30 to 40% of the enlisted men, and one-fourth of all first termers have been married; and the military has responded appropriately with increased dependent allowances and a full range of family community services (Little, 1971; Moskos, 1970; Steiger, 1971). Further, in recognizing that it cannot afford to compete with the family for a soldier's loyalty, the military has made a concerted effort to bring the enlisted family increasingly within the military system through on-post housing and other programs (W. Bennett et al., 1974).

The case for continued and even greater consideration of the family in military policy making and allocation of funds is a convincing one. First, it stands to reason that a contented family life will bolster a soldier's morale, while men who are experiencing family problems will have lowered efficiency on the job. Christie (1954) found, for example, that Army recruits who were married adjusted to military service much better when given greater contact with family and friends. It is also inefficient to train men for a job and have them leave the service, and it is estimated that many men

AUTHOR'S NOTE: *Appreciation is extended to Nancy Pollock for assistance in the literature review and the preparation of this paper.*

do so because of an inability to arrive at a satisfactory family adjustment within the military context (Coates and Pellegrin, 1965). In fact, Vineberg and Taylor (1972), in a study of the Modern Volunteer Army program, found a number of family-related variables to be important for personnel considering reenlistment, while A. Owens (1968) also found family responsibilities and influences to be important in Australian Army reenlistments. The importance of the family for AWOLs was demonstrated by Hartnagel (1974), since over half the AWOLs he studied were "family problem solvers" who had gone AWOL in order to correct family problems or to alleviate family-related financial difficulties. Wives' attitudes are also important, as shown, for example, in a study by Clum and Hoiberg (1971) of Navy and Marine Corps psychiatric inpatients in which the attitude of a patient's wife toward continuance in the military was an important factor in his ability to readjust to the demands of the military. Further, since 89% of all Army jobs can be filled by women and approximately 16% of the ROTC force is female (Blonston, 1974), it is likely that more families will emerge in which either both spouses are on active duty or a husband qualifies as a military dependent—trends which certainly have implications for the importance of the family in military policy. Finally, throughout my own military service I often heard the sentiment expressed, especially by careerists, that the Vietnam War was justified on the grounds that "I've got a wife and kids and I'd rather be fighting over there to protect them than wait 'til the enemy reaches our shores." While this is obviously not the only motivation for a service career, I would posit that it is a factor in the desire for military engagement and justification for war and, at the very least, attests to the protectiveness and importance that many military men feel for and attach to their families.

Societal Trends and the Military Family

The implications of overall societal trends for the military family—which is, of course, a microcosm of the family in the larger society—are considerable. Problem areas such as family mobility and retirement are of particular relevance for the military.

A number of authors, including Dickerson and Arthur (1965), Janowitz (1960), and Little (1971), have noted how military life has loosened ties between servicemen's conjugal families and their extended families and communities. In a sense the military was a precursor of this trend in society as a whole. Should the trend level off overall, however, there are few indications that the military would follow suit and locate families closer to their homes of origin. It most certainly will not change direction unless this practice clearly becomes detrimental to retention in the service or unless a concerted countermovement develops among military personnel themselves.

The military has definitely not been untouched by changes in women's roles and independence. An increasing number of servicemen's wives have taken jobs to supplement their income (Steiger, 1971). Since this trend emerged primarily in response to the low level of servicemen's pay (contrasted with economic inflation and demands for a higher standard of living), it may level off somewhat with the substantial pay increases granted the all-volunteer force. Nonetheless, it is likely that women serving as military wives, as active-duty personnel themselves, or in both roles will become even more influential in military policy. It will not serve the military well to ignore them; they affect retention in both the present and future. An example of this was provided 15 years ago when Janowitz (1960), in his interviews with career officers' wives, found that, despite their professed allegiance to the military, they were much more vague and tentative than their husbands when asked whether they wanted their sons to be officers.

It will be difficult for the military, should it so choose, to bridge the differences which have evolved between older and younger generations and to reverse the greater dependence upon their peers that has emerged among youth (Bronfenbrenner, 1974; Ryder, 1974). Efforts have been made to improve the social lives of dependent children through teen clubs, Boy Scouts, and other on-post, military-sponsored activities. (Although officers and their children tend to favor some of these activities, enlisted men and their offspring favor others [Little, 1971].) The issue, however, is larger than the simple availability of social and recreational resources, since factors such as family mobility, father absence, and an increase in more socially conscious education are important. Further, many would question whether it is desirable to attempt to reverse some of these trends. Of course, recent times have seen an apparent narrowing of the generation gap, and the schism may all but disappear within the next generation or two.

Rossi's prediction (1972) that families will become more self-examining and open to change is an important one to consider. I see this happening now when I examine the news media, professional literature, the content of seminars at professional conferences, and my own clinical experience. From the standpoint of the military, it could have both positive and negative implications. If families could take it upon themselves to develop a number of ways of coping with their own interpersonal or military-related problems, it would be useful. On the other hand, greater egalitarianism and a propensity for questioning values and life goals could be seen as quite threatening to an institution as historically tradition-bound, hierarchical, and (frequently) rigid as the military.

THE CIVIL-MILITARY INTERFACE

The relation between civilian and military structures has obvious family implications. Although it is too large a topic for more than superficial coverage here, certain aspects do merit discussion, since a number of military policies, practices, and programs have recently emerged which affect this interface at the family level. Little (1971) notes that, through programs such as the Civilian Health and Medical Program, United States (CHAMPUS), the military family has grown less dependent on service facilities, and he predicts that such programs will become increasingly significant in the civilian community. Also, as education for officers and other servicemen has increased, there has been a growing resentment toward isolation in the "khaki ghetto" and a desire by many of them to lead more "normal" lives (Harwood, 1970). Further, the greater exposure to television and the mass media may have fostered this desire by reminding military personnel of the way that others in society live, thereby countering what Janowitz (1960) has identified as the customary tendency among the military to avoid making comparisons with civilian standards of living. Another example has been the policy instituted by General Creighton Abrams in 1972 of toughening up training and insisting on discipline during duty hours while allowing more civilian-type freedoms during off-duty hours, thus in some ways making military life more comparable to civilian life and possibly fostering joint civilian-military recreational contacts and activities. Finally, Moskos (1973a) foresees a pluralistic or segmented trend in the military in which the combat arms would grow more divergent from civilian society, while military technical support agencies would grow more similar; the implication is that military *families* would likewise differ in their resemblance to and perhaps their contact with civilian society.

In contrast to the above, a number of factors can be identified which may contribute to a divergence between the military and civilian spheres and their inclusive families. Gitter (1973) notes that during peacetime the serviceman becomes bored, feels unappreciated and isolated, and tends to increase his reliance on tradition and authority. Gitter describes the apartness which this breeds and offers as an antidote such programs as the Department of Defense's Domestic Action Program (which he laments is presently not directing sufficient effort to helping those within the civilian sector who are in urgent need). Finally, Moskos (1970) and D. Segal et al. (1974) believe that expanded on-post housing, in conjunction with other programs designed to bring the pattern-maintaining functions of the family within the military system, will reduce civil-military interdependence at the individual and family level, resulting in a more closed system. It is not possible at present to assess the extent to which these developments will offset the factors mentioned in the previous paragraph, but in view of the events of recent years, it seems safe to assume that the civilian society will exert considerable influence on all facets of military life for some time to come.

Family Mobility

Relative to the larger society and particularly the families within it, the military has been, in the literal sense, a prime mover. Frequent, often unexpected, relocations have long been characteristic of military life, especially during wartime. This practice is often inefficient and usually costly to the military (Stanton, 1973), and it has been such a problem for military families that it was a major reason for instituting the Army Community Service and the Air Force Family Service programs. Because of it, Frances and Gale (1973) note that military family members try to lessen the hurt of separation by avoiding deeply felt extrafamilial attachments or, conversely, in some cases, by engaging in intense, hedonistic relationships which are short-lived. Little (1971) has noted how residential instability can interrupt long-term medical care, children's schooling, etc., but he and others also feel that there are benefits such as a wider range of experience for children, the inducement for neighbors to help each other, and the facilitation of family coping with immediately unpleasant situations through the expectation that such conditions will not have to be endured very long.

A good deal has been written about the effects of residential instability on the family, much of it from a mental health perspective. McKain (1969) found that mobility was particularly stressful for wives who were alienated and did not identify with the military—especially those who lived off post. Pedersen and Sullivan (1964) uncovered no differences in frequency of geographical relocation between "normal" children and child clinic cases, although attitudes of parents of normals were more positive toward the military (both parents) and toward mobility (mothers) than were parents of disturbed children. Shaw and Pangman (1975) corroborated these frequency-of-mobility findings; however, they also noted selected instances in which the family move had an obvious adverse effect on a child's emotional well-being, and they underscored the difficulties that a move can engender in an adolescent who, because of it, is forced to become more dependent on his parents at a time when he is struggling for independence. Along these lines, both Darnauer (1969) and Gabower (1960) found that adolescent military dependents were more negative toward moving than younger children, primarily because their peer-group relationships were of greater importance to them. Further, a study of adolescent boys by Wooster and Harris (1972) found that those whose families were highly mobile showed a deprived self-concept, while Lyon (1967) has presented evidence that frequent family mobility has a deleterious effect on both standardized test performance and psychological adjustment in children of U.S. Marine Corps personnel.

A special case of family relocation is assignment to a foreign station. Little (1971) notes that such assignments are considered highly desirable. The reasons for this are complex, but he cites economic factors and the solidarity among service families living abroad as of paramount importance. There are also a number of drawbacks, among which McNeil and Zondervan (1971) include culture shock, perceptions of living a tremendous distance from home, boredom, changed life style, and, primarily owing to isolation, an inordinate dependence of wives on their husbands. In any case, prospects for the foreseeable future are that the number of overseas assignments will be dwindling and will be available primarily to those who request them, so that this issue will probably decline in importance.

While a certain amount of family mobility will always be a feature of military life, it appears that pressures will continue to mount against its prodigiousness. The increasing influence of wives, along with greater recognition of the importance of family elements, will serve to temper this practice and bring it more in line with civilian norms.

FAMILY SEPARATIONS AND DISRUPTIONS

Related to family mobility are the separations and disruptions frequently experienced by military families. As with relocations, these generally result from reassignment of the active-duty member—usually the father—so that, even during the relative stability of peacetime, 15 to 20% of married servicemen do not live with their families (Moskos, 1970). The problem is particularly acute for husbands assigned to tactical or combat-ready units who are also on call much of the time and unable to establish regular family routines. At such times, wives are under particular stress and must be prepared to assume both parent roles, a pattern which fosters matriarchal family structures (Lindquist, 1952; Little, 1971). While it is probably true that a military career may serve to legitimize an unconscious desire by some couples to actually be separated (Frances and Gale, 1973) and that the effects of separation are greatest on families which already show a certain degree of instability (Boulding, 1950), studies of servicemen's children (Baker et al., 1967), Army and Air Force wives (MacIntosh, 1968), Navy submariners' wives (Pearlman, 1970), and POW's families (McCubbin et al., 1975) indicate that the more "normal" families do not always escape father-husband absence unscathed.

What happens to families during these separations? Most often they are required to move off post and relinquish ties to the military community when these ties are most needed, especially if relatives are not close by to provide support. If such relatives are available, excessive dependence on them may develop. Infidelity (or at least exaggerated fears of it) may occur, sometimes leading to divorce (Lindquist, 1952). During crises, families without nearby relatives tend to turn for support to close neighbors, chaplains, and, almost as a last resort, services within the formal military community (Allen, 1972). Montalvo (1968) showed that such families failed to adequately utilize civilian resources, and he made a case for retaining these families within the military community so that its support system could serve as protection against the stresses of family dismemberment.

The military shares the dubious honor with a few other occupational groups and institutions of being a pioneer in the trend toward the parental absence that has emerged in our society. Forced family separations are probably a major reason that many servicemen do not reenlist (Coates and Pellegrin, 1965), but the prospects for altering military assignment policies are not good, particularly for those in tactical units. Should society act to counter the parental absence trend—and there are indications that it may—it remains to be seen if the military will follow suit.

RETIREMENT AND THE FAMILY

Nearly every career military person faces retirement and a dramatic change in job and environment earlier than his counterparts in the civilian work force. By 1980 there will be a projected one million persons in society, with their attendant families, who will have experienced the difficulties of retiring between the ages of 40 and 50—a time when most civilians are reaching the peak of their earnings and productivity (Giffen and McNeil, 1967). Whereas the retirees used to cluster around military bases, CHAMPUS has permitted them to disperse, so they are coming more in contact with nonmilitary citizens (Little, 1971). As they enter society at the rate of 50,000 per year (Bellino, 1969), the retirees and their families will put increasing demands on CHAMPUS health care and other retirement programs. Unless such programs are transferred to the Veterans Administration or bureaucratically handled in some other way, they will take a bigger bite of defense spending. In addition, since a greater proportion of the members of the all-volunteer military force will be career personnel, a greater proportion of all military will face the early retirement experience. Consequently, these retirees will have a greater impact on both the military and the civilian spheres. That more servicemen and their families will share the experience of and concern over retirement means that it will become a more salient concern of military policy makers—and one of the big worries of military retirement is family security. Further, the increasing number of retirees, being less dependent on military bases and more civilianized, may be more likely to adapt to the values of the civilian world when these conflict with military values. To the extent that they accept the predominant civilian views of the family and to the extent that, in their increasing numbers, they affect military policy, to the same extent will the civilian world exert influence over the military's view of the family. In such roundabout ways, then, military retirement will have implications for the future of the military family.

New Directions

A number of policy changes and programs will be proposed in this section aimed at resolving some of the differences in priorities that exist between families and the military institution to which they are tied. Although the military has been innovative in developing programs that administer to its families, there is room for improvement. Some of these programs may serve to separate the military and civilian spheres further or at least to establish intramilitary structures which parallel those of the civilian world. However, in most cases it is felt that this will be necessary in order to facilitate the coping and adaptive functions of families in the military, heighten the contributions that these families make to military effectiveness and military life, and reduce the possibility that the all-volunteer force will be primarily composed of individuals who either do not or cannot maintain families of their own.

THE FAMILY AS THE PRIMARY UNIT

Although the military has increased its family emphasis in the allocation of resources, assignments to installations, and other personnel or management activities, the basic unit it deals with is the individual serviceman. I propose that this be changed. I suggest instead that *families* serve as the primary units for management decisions. While in one sense radical, this idea is in other ways not such an extreme departure from conventional views, since the military customarily thinks in terms of such other multiperson units as platoons and squadrons. With such a system, the units would, of course, vary individually with number of dependents–unmarried personnel would be one-person units–but the concept would be manageable, given actuarial data on family size for various ranks and specialties. In addition, such a schema would also help to bring the view of the family as a system (McNeil and Zondervan, 1971) in line with other systems concepts used at various levels of management.

DEPARTMENT OF DEFENSE FAMILY AGENCY

There is thus a need for greater input on family issues at the highest levels of Department of Defense policy making. Although such agencies as the Army Community Service have served this need to some extent, their responsibilities are diluted with less family-oriented concerns; they are often more concerned with "pots and pans" problems, and, it seems, they do not have the necessary clout to be influential in key policy decisions. What is needed is an agency, analogous to the Department of Marriage and the Family proposed for the presidential cabinet level by Rue (1973), which exerts influence at the Department of Defense level in the areas of family health, adaptation, welfare, etc. An example of one of its functions might be to require a family impact statement, much like that proposed by Senator Mondale for other arms of the government, which would force Defense agencies to "discuss and debate the possible consequences of a given policy on the family."

PROGRAMS AND POLICIES

Over the last few decades, the family in general has become more important to the military, and specifically, as Little (1971) points out, the family of each serviceman has become more important in each step of his military career. The host of family health, economic, and other programs implemented by the military have aided retention, brought comfort and continuity across the many military installations, and served as models for the development of similar programs in many facets of the civilized world. In the fields of equal (nondiscriminatory) employment and comprehensive health care, the military has been a pacesetter. It would therefore be unfortunate if the military were to rest on its laurels and not continue to seek new and better ways of serving its families and thus increasing their contribution to its mission. Family needs of the future require extension beyond simple economic and health matters and include aiding families in coping with what General Robert Gard (1974) has termed the "uncoordinated organizational maze" of the various helping services, plus considerations noted by Rue (1973:690) for guaranteeing "growth, maturity, satisfaction and a mentally healthy environment for marriage and family life."

In the succeeding paragraphs, areas will be highlighted in which changes in emphasis or direction of existent programs and policies are called for. The list is at most exemplary and by no means exhaustive. While it is recognized that some of these proposals, if adopted, would serve to bring families of servicemen even more within the military system, it is felt that they are most appropriately implemented in this way, since most of them would be difficult to manage otherwise. Perhaps a greater danger would be that they might be differentially applied to the combat versus the technical-support agencies, a move which, in addition to leading to greater segmentation, would be unfair and contribute to interagency antagonism.

REDUCED FAMILY DISRUPTIONS

Clearly, problems of relocation and parental absence are among the greatest the military family has had to endure. Although the Department of Defense appears to be attending to these disruptions more within the all-volunteer context, one can question whether present changes go far enough or in the most appropriate direction. Concerning mobility, for example, I have not learned of any programs which take the one group most adversely affected by it into account, i.e., adolescent dependents. The number and extent of negative family reactions to relocation would surely be reduced if Defense policy were to minimize reassignments for families with adolescents. Further, part of the computer data base for a particular serviceman and his family could be an annually updated assessment of the family's favorable or unfavorable attitudes toward relocation. Such a system would be flexible enough to take changes in attitude into account and might help to reduce the sense of powerlessness so often felt by people in the military.

During the absence of a parent, the family redefines itself. The remaining parent becomes head of the household, assumes more responsibility, and becomes more independent. When the absent parent returns, the sparks often fly and both spouses must renegotiate their roles. Often families which adjust best to the absence of a parent have the most difficulty readjusting when that parent returns (McCubbin et al., 1975). A reorientation course for servicemen and their wives given prior to and just after return is one way the military could alleviate the pain of readjustment. Such a course could help them adapt and could increase the servicemen's efficiency during this critical period.

FAMILY GROWTH AND MENTAL HEALTH PROGRAMS

The expansion of CHAMPUS to cover outpatient psychiatric treatment, as well as the continuing shortage of psychiatrists and other mental health professionals within the military, has served to make the problems of military people of more interest to civilian psychotherapists (MacIntosh, 1968). Nonetheless, the military has not and cannot abdicate responsibility for the psychological well-being of those within its fold. There are a number of areas in which changes and expansion of programs and policies toward family growth and mental health are in order. Some major ones are presented below.

Family therapy. Although Bellino (1969) and Pearlman (1970), among others, have advocated its application to military personnel, the family approach to dealing with mental health problems has not been widely practiced within the military. This is unfortunate, for family therapy has much to offer those who face the kinds of problems discussed previously. While in many ways consonant with other psychotherapeutic approaches (Stanton, 1975), it is usually briefer than most individual therapy. It is an approach which can be especially effective in treating alcoholism and also in bringing into treatment people, such as parents of an unmanageable child, who need help and would not otherwise seek it. This form of treatment, then, closely adheres to the needs of the military family and should be more extensively adopted.

Separation of treatment from job functions. Although increased on-post housing and services function to bring the on-post military family under greater social control, the practice of reporting to a serviceman's commanding officer any untoward activities, such as family quarrels, delinquent children, etc., which occur outside his military duties often hampers the individual's promotion possibilities and discriminates in favor of those living off post, since reports of the latter's problems rarely reach their commanders. Such information is usually conveyed through military police "Delinquency Reports," reports by medical or mental health authorities, or even gossip among neighbors and peers (Little, 1971).

Related to the above are the issues of confidentiality and privileged communication within the mental health sphere. Military personnel are not afforded this luxury of confidentiality. Mental health confidentiality is practically nonexistent in the military, and commanders can obtain what records they want without a great deal of difficulty (Stanton, 1973). For this reason, it is not uncommon for servicemen (especially officers) to seek help surreptitiously from private civilian professionals when needed, in spite of the availability of both CHAMPUS and military mental health services. Moreover, personnel on isolated posts may not even have this nonmilitary option open to them. Justification for nonconfidentiality is usually based on "security" needs. However, even where this applies, it is not always necessary for the commander to be informed.

Family outreach and growth programs. A major conclusion emanating from the research by McKain (1969) was that during times of stress military wives, especially those who feel alienated, are not contacted frequently enough and provided with the supports that exist within the military community. In addition to suggesting that such wives be moved on post, McKain recommended "reaching out" programs as a means for improving this situation, using, in particular, peers and neighbors as contact persons. The work of McKain and others evinces the extent to which the majority of military family services are either recreational, ameliorative, or of the "pots and pans" variety. Little attention has been given to programs directed toward growth and prevention of problems. One possibility to improve the quality of family life would be an adult education program for couples and, at times, their children. The curriculum might include (a) methods for dealing with distinctive aspects of military family life and (b) methods of coping with experiences that most families encounter, such as conflict resolution, premarital issues, the "empty nest" syndrome, financial management, and communications skills (e.g., active listening, clarifying expectations). Such a program should be action-oriented, educative, and nonjudgmental and should emphasize growth rather than "problems." This approach would be, in every way, a preventive one and aimed at inoculating families against future difficulties in addition to providing them with techniques for bettering their lives and relationships.

Implementation of programs such as the above would obviously be limited by the number of available military personnel in community service, mental health, and other related disciplines. One method of dealing with this limitation is the aforementioned tactic of enlisting selected peers and neighbors or neighborhood leaders for specialized training and letting them shoulder much of the program responsibility. Military community services have done this sort of thing for a number of years through volunteer programs.

Retirement programs. Retirement is a time of crisis for the serviceman and his family, and psychosomatic problems are not uncommon in both the active-duty member and his dependents during this period (Giffen and McNeil, 1967). In recent years the military has felt an obligation to help its people through this process, and the creation of Project Transition is one manifestation of this concern. Project Transition is primarily oriented toward vocational readjustment and training, however, and does not meet other major needs of the retiree. Additional programs are required to help with some of the emotional aspects of retirement, and such programs should not only include the active-duty person but, as Bellino (1969) recommends, his entire family also. Further, Giffen and McNeil submit that service programs of this type should start at least a year or two before retirement, because that is when problems frequently begin.

In view of the above, it is proposed that a counseling and educative program be instituted within the military geared toward mobilizing the strengths of the family for coping with retirement. It should commence well before the date of discharge and extend into civilian life if need be, since the problems that are likely to arise are of a service-connected nature. It should be managed either collaboratively by Community Services, Project Transition, and mental hygiene facilities or by a separate agency. It might include a prevention-oriented, preparatory short course for retirees plus ameliorative services. The training of military professionals in military retirement counseling would be an integral part of it. It might also be extended into the civilian community by providing local nonmilitary professionals with short courses or workshops directed toward dealing with the problems of military retirement. In short, the proposal is for a program which is longitudinal, has preventive and ameliorative components, and draws on the resources of both the military and civilian societies.

Summary

In this paper, trends within the military family have been discussed in relation to its past and future, and also in relation to families within the larger society. Military families are important to the military, and a number of programs and policy changes directed toward the military family have been proposed in such areas as family treatment and growth, the retirement process, family disruptions, and consideration of the family as the primary unit for management decisions. While some of these changes might serve to further insulate military families from the civilian world, it is felt that ongoing changes in the larger society and in the civil-military interface will tend to offset such a trend.

THE ARMY OFFICER'S WIFE: SOCIAL STRESSES IN A COMPLEMENTARY ROLE

Ellwyn R. Stoddard and
Claude E. Cabanillas

Those who choose to become Army officers aspire to success just the same as those engaging in any other kind of work. However, in few jobs do the symbols of success seem as visable as the attainment of rank in the military structure. To succeed, the officer must perform well in his assigned duties, and he must also have the unquestioned support of his wife if he is to reach the highest Army positions. For her, sharing in her husband's occupation while having to maintain her familial and personal responsibilities produces additional pressures and frustrations which are not well understood by her officer husband in the male military fraternity.

The aim of this research is to clarify the role of the Army officer's wife and to show the changes occurring within it. We hope to document the role stresses which are experienced by the Army officer's wife and relate these to the various stages of her officer husband's career. The guiding hypothesis suggests that role stresses for the officer's wife will increase directly with the number of years of her husband's service and with his attainment of higher military ranks throughout his military career.

Background Perspectives

Current surveys of military research completed by social scientists reveal that the bulk of it pertains to military organization and is focused on military personnel themselves. The remaining studies deal with policy matters, political influences in military functions,

AUTHORS' NOTE: *Dr. Ellwyn R. Stoddard, currently a professor of sociology and anthropology and formerly an artillery officer, and Captain Claude E. Cabanillas, on active duty in the Signal Corps, on May 1, 1975, presented an abbreviated version of these research data at a meeting of the Western Social Science Association at Denver, Colorado. The views and comments advanced by Captain Cabanillas in this article are his own and are not those of the Signal Corps, the U.S. Army, or the U.S. Army Sergeants Major Academy.*

technology, and its impacts on the military establishment, and similar studies of random substance and varied scope. Beyond the peripheral role of the family in the effective functioning of military personnel, little notice has been given to the military family per se (Lang, 1972; Kourvetaris and Dobratz, 1975). One reason for this paucity of material may be historically the military policies of ignoring family life and financial compensations for the married soldier until the period of World War II. Thereafter, the heavy influx of family men entering the military service required some planning of allotments, compensations, and services at Army installations for wives and children (Little, 1971). But with these policy and expenditure changes, there was little or no awareness of the changes in functions, requirements, and distinct background origins of current Army officers' wives. The expectations of their involvement in the officers' careers are based on the pre-World War II codes, which were mostly applicable to senior officers, who were the only ones to enjoy an on-base family.

The written materials which are currently available describing the Army officer's wife and family are generally of three types: treatises on earlier policies concerning military families; personal narratives of military experiences and reactions; and sophisticated descriptions of the ideal prescriptive models of military family life as reflected in military publications, pamphlets, and wife guides (V. Johnson, 1967; Little, 1971; Lindquist, 1952; Janowitz, 1971a; Shea, 1954; Murphy and Parker, 1966; Kinzer and Leach, 1968). Thus, inasmuch as the U.S. Army currently follows the policy of transferring officers as individuals rather than as total units from time to time throughout their careers (Janowitz and Little, 1965:82), the military family is quite likely to be the only social unit with which the officer will associate himself throughout the whole of his career, making family support one of the most important factors influencing effective performance of the Army officer himself.

The Army's lack of concern for research dealing with its military families indicates a high level of successful performance of this institution in the past. But gradually the instability of marriages and family disintegration so very evident in civilian life is becoming common to military personnel as well. In addition to this, the growing influence of the Women's Liberation Movement in trying to destroy the demeaning female stereotypes of the past and give a new positive image of independence and self-respect is not lost on many officers' wives, who review their own circumstances in the light of this new perspective. It could well be that the future strains within the Army's organization will not be those of race, ethnicity, and sex integration of military personnel on the job, but rather those developing in the homes of military personnel between husbands and wives.

WIVES' INVOLVEMENT IN HUSBANDS' CAREERS

Social science literature dealing with the relationship of wife functions in the career of the employed husband might be more easily understood if categorized into three major orientations. The first of these, represented by "the housewife," features a total isolation of the wife from her husband's work. This is programmed not by the conditions of the occupation or the employing organization but by default of the wife as a result of her lack of personal interest or her husband's policy of noninterference. The second type, characterized by the "corporate executive's wife," contains patterns of wife isolation or involvement in the husband's career extending the complete range of the spectrum. Unlike the isolation of the first type of wife, however, her involvement or noninvolvement in her husband's occupational success is regulated by her husband's employer; it is not a function of domestic policy between the spouses. The third type, characterized by diplomatic career officials, some clergymen, and other professionals, prescribes that the wife fully participate in the duties and activities surrounding the husband's formal appointment or occupation as a partner or complement to his career successes. Her involvement is not regarded as a potential interference with the husband's commitment to his work.

Thus, in prescribing varied patterns of wife involvement with the work career of her husband, social scientists have outlined at least three major typologies ranging from complete isolation to partial integration into the husband's occupational pursuits. These patterns are represented by the *housewife,* the *executive's companion* and the *complementary wife.*

The familiar *housewife* role, described in greatest detail in the study of Helena Znaniecki Lopata (1971) ascribes to the spouses a polarized division of labor: the housework and child-rearing functions are the exclusive property of the wife, and the world of work is the sole obligation of the husband. Once the husband leaves for the job, he is "out of sight, out of mind" until he enters the house at the conclusion of the working day. He also pays minimal attention to his familial duties.

The *executive's companion* is a company asset, who works for her husband's successful advancement within the corporation but does not attempt to influence directly the important work of the corporate structure. The company either isolates her from company business or successfully enters into a compact with her, seeking her influence in making him dedicate his full energies to the goals of the corporation without dissipating his time with home responsibilities. As described by William H. Whyte (1956), her major duties–to remain in the background and to come forth and exhibit her companionate skills when needed–will keep her husband happy at home and will result in better quality and more quantity efforts from her husband during his career.

The *complementary* (or vicarious) *wife* is best illustrated by the ambassador's wife studied by Arlie R. Hochschild (1969). She, in her quasi-legal capacity must share and supplement the official duties of her husband–in this case, by communicating symbolically the correct relationships between their government and others, mainly by engaging in the subtleties of diplomatic life and reciprocal entertaining. Although she, like the other wives discussed, functions as an ancillary but involved extension of her male spouse, her active participation in the career management of her husband's ambassadorial appointment shows a far greater wife involvement in the husband's occupation than does the corporate executive's companion or the housewife.

THE ARMY OFFICER'S WIFE

Although elements of the housewife role, the executive's companion role, and the complementary role are present in the role of the Army officer's wife, a more detailed examination of her duties, functions, and role prescriptions will show which of the above three female types is the model closest to approximating that of the Army officer's wife.

A superficial glance at the elitist corporate executive stratum and its equivalent among Army officers might yield the impression that they are extremely similar–the portrayers of an elite life style within an organization that requires deference to them from subordinates, concentrates great power in their positions, and has adequate procedures to maintain them in their exalted position. Neo-Marxist writers have developed this theme to a full-blown power conspiracy of corporate, political, social, and military elites in America (Mills, 1956; Keller, 1963).[1] But upon a more detailed examination, the many differences between corporate and military policies, personnel, and occupational prescriptions become evident.

At the present time the military differs markedly from its corporate counterparts in the scope of company-sponsored benefits. Whereas the corporate structure seems willing to provide well for its management personnel but not for its massive numbers of production line workers, the military has become dedicated to the task of devoting a greater and greater portion of its budget to benefits such as housing, education, etc. for its enlisted personnel (D. Segal et al., 1974; D. Segal, 1975b). But an even greater difference between these elites in private and public sector institutions arises from the distinct historical legacy and functional goals of the two. The physical and social isolation of the military from civilian society is one of the major differences noted.

> The intimate social solidarity of the military profession, which civilians often both envy and resent, is grounded in a peculiar occupational fact. Separation between place of work and place of residence, characteristic of urban occupations, is absent. Instead the military community is a relatively closed community. . . . The sharp segregation between work and private life has been minimized in the military occupation. [Janowitz, 1971b:177-178]

A further source of civilian-military officer alienation has been the rigorous training and special socialization of military cadets who ultimately become the senior officers of the Army. The formal and informal procedures of this distinct culture and the powerful in-group loyalties are developed and/or amplified within the Academy or ROTC corps of cadets (Dornbusch, 1955; Picou and Nyberg, 1975). At odds with this view is J.P. Lovell (1964). The traditional regional and class origins of senior Army officers have been maintained, with occupational inheritance and endogamy by means of selecting marriage partners from among the daughters of the military community (Janowitz, 1960).

Other than the common role stresses accompanying the domestic duties and familial roles endemic to the occupation of housewife, the military wife shares little of the housewife's insulation from her husband's occupation and career (Lopata, 1971). The military wife knows of her expected involvement in military and community affairs as an adjunct to her officer husband's quasi-formal responsibilities.

Another major factor which distinguishes the military occupations from most others within the civilian work force is the unique nature of the work contract and the unlimited time "on call" which can be demanded in wartime without additional compensation, as compared to the 40-hour week with overtime in the civilian sector. A succinct commentary by David R. Segal (1975b:2-3) further explains this factor of unlimited service:

> The soldier, unlike his civilian counterpart, enters into a contract of unlimited liability with his employer. He cannot unilaterally terminate his employment at any time he wishes during his period of service. . . . He is frequently called upon to work more than an eight-hour day, for which he receives no additional compensation. And in time of war, he receives no additional compensation. And in time of war, he must face prolonged danger and may even forfeit his life. Obviously the man on the firing line is required to make commitments of a different order from the man on the assembly line.

In addition, the element of mobility distinguishes the military officer from men in most other occupations. Although some business executives as well as military officers are subject to transfer within the organization, necessitating the creation of a new social milieu, the routinized pattern of 3- to 4-year tours for the Army officer produces a great deal more movement throughout a career than that experienced by most business executives. However, the disruptive element of relocation ascribed to the wife of the business executive is not wholly present among Army officers' wives. Rather, when the current tour nears an end, there is a psychological preparation for discontinuing one's social responsibilities in the present location and beginning to speculate about one's future assignment. When officers receive their orders to transfer to another station, those officer couples who arrived on post about the same time become very uneasy until their orders (or some informal word concerning their next assignments) come through. In this, then, both the officer and his wife share the anticipation and preparation which apparently are more of a strict male decision for the corporate executive and his family as stressed by Seidenberg (1973).

The pattern of frequent dislocation tends to dull the sensitivities of military adults toward the civilian community and its problems. For the Army officer and his wife, there are immediate demands to become involved within the military community and the wives' organization, but they are often dragged reluctantly into community policy and problem issues, civilian voluntary organizations, and local bond elections. However, with the months necessary to establish new social relationships and to "know the territory" combined with the months at the end of the tour when psychological alienation accompanies anticipation of the forthcoming move, there is very little time to become more than superficially involved in the larger community.

In light of the traditional military isolation from civilian society, the peculiar occupational demands of military life, and the patterns of frequent physical relocation, the military has preserved a distinct culture of its own, complete with its own version of stresses, strains, and maladies. But if this historical isolation of the military from the larger society is decreasing–as most authorities believe it has been since the big changes in World War II (Biderman and Sharp, 1968; Moskos, 1973a; D. Segal et al., 1974; Sarkesian and Taylor, 1975)–it may be anticipated that the liaison between military and civilian voluntary organizations, service-related projects for community betterment, and social intercourse will be primarily the function of the officers' wives. It would require more and more out-group contact with nonmilitary couples and groups, and these activities might legitimate excuses for nonconformity in the strictly military sphere; the stresses and tensions accompanying their civilian liaisons might well reflect those in the civilian sector rather than those of traditional military families.

The socialization of an Army officer's wife follows quite closely that of the ambassador's wife. As in embassy service, the senior officers' wives acquaint the newcomer with the hazards of her role. She is told what her days as an officer's helpmate should be like and that if she is to uphold the military traditions and be a complement to her husband officer, military considerations should have priority over personal interests, friends, and nonmilitary loyalties. Written guidebooks and pamphlets list specific duties and obligations that she is to assume and strongly suggest her involvement with other officers'

Table 1. MARITAL STATUS OF ARMY OFFICERS BY GRADE (%)

	Col.	Lt. Col.	Major	Captain	1st Lt.	2d Lt.	WO
Married	94.8	95.5	93.0	85.2	69.3	58.8	89.5
Single	5.2	4.5	7.0	14.8	30.7	41.2	10.5

SOURCE: Office of Personnel Operations, Personnel Management Development Office, Department of the Army DAPO-PMP Report No. 42-72-E (samples as of 31 August 1972).

wives in their organizations (Kinzer and Leach, 1968). For those wives who realistically covet the very top positions within the Army for their officer husbands, the criteria for general officers outlined a few years ago by Robert F. Froehlke (1972), Secretary of the Army, make them even more conscious of their role in their husbands' future career. Froehlke explicitly mentioned that an officer being considered for potential senior appointment should be rated negatively if his wife would be a potential embarrassment to the military. On the other hand, if there is no wife or if the wife is a burden because of infirmities, he should not be considered as having negative attributes.[2]

Such compensations as dependent allowances, household transfer costs, housing allowances, and medical outpatient services have increased during the past three decades and have served as an enticement or subsidy for the married man with a family. But it may well be that a wife *is* a requisite for the success of an Army officer, and for this reason, more than any other, most officers are married, as revealed by the data in Table 1.

Traditionally, Academy cadets have postponed their marriage plans until after their commissions as officers have been received. And then for them to choose a wife from a well-known military family has been no handicap to high visibility among the senior military brass. On the other hand, the huge numbers of ROTC and OCS officers brought into the military service during World War II and the Korean War and again during the Vietnam conflict had often been married prior to their officer training. And the wives of these non-Academy officers are more likely to have developed loyalties and interests divergent from those rigidly specified for the traditional officer's wife. In this sense, there has been an increasing rate of nonparticipation in strictly military affairs by the modern Army officer's wife, who is a converted civilian with civilian tastes and loyalties at odds with the stringent boundaries formerly placed around an aspiring officer's wife.

In summary, although the Army officer's wife shares many of the same stresses experienced by the housewife and the executive's companion wife, her prescribed role is much more closely tied to the complementary wife model. Employing this as a background against which to examine a specific sample of Army officers' wives, we will now examine the specific stresses within their roles with which they must deal.

Theoretical Considerations

For a more thorough analysis of role stresses experienced by the Army officer's wife, some conceptual tools are required. Role theory provides us with some useful tools explaining role incongruity and dissonance, but, inasmuch as their definitions are not altogether standardized and they are often employed with imprecise meanings, it is necessary to define and clarify them here. There are occupational studies dealing with this subject that are applicable to military roles (Turner, 1947; Burchard, 1954; Stoddard, 1954). Role theory posits each individual as a total social person whose links to many diverse groups and individuals in distinct capacities or mutual obligations (roles) create a rather complex set of expectations within each individual.

Inasmuch as the complex stresses within the category of role theory are so varied, these must be reduced further to more specific role concepts. In our research we have concentrated upon four of these. *Role conflict* and *role strain* are central to the focus of the study, while the related stresses of *role ambivalence* and *role alienation,* though somewhat periperhal, are employed from time to time.

Role conflict. This is a form of role stress which occurs *between social roles* carried by a single individual, wherein the expectations of one role preclude the successful performance of the other. This may come about because of limited resources, such as having to be at two locations at the same time or having to expend unusual amounts of energy and money for one project which should be reserved or have been promised to another. The selection of one of these would automatically exclude the possibility of fulfilling the expectations of the other role and its requirements. Thus the Army wife's involvement in her husband's career gives a firm priority to Army matters, ceremonies, activities, and responsibilities over her personal interests and familial roles, and this to a greater extent than the normal housewife, who is essentially alienated from her husband's work except for rare social engagements which even then may be avoided if other matters are pressing.

Role strain. This indicates a stress situation in which obligations *within a given social role* are in competition for the time, energy, and financial resources of a person—e.g., informal "local traditions" which take unauthorized license with the regulations. Role strain is less severe a problem than role conflict, although it may be a more common occurrence, owing to the many dissonant aspects of each social role.

Role alienation. This occurs when a person no longer feels compelled to fulfill his social obligations and withdraws from the source of role expectations for which he no longer feels responsibility, a cop-out as it were. It may happen that his duties reach a saturation point, and he cannot tolerate them further; and, in lieu of direct conflict, he engages in a psychological blocking of further duties, often by becoming estranged or alienated from the source of the problem.

Role ambivalence. This exists when obligations incurred exceed the resources available, and there is no hierarchical ordering of priorities in which to assign the available resources of time, personal energy, and money; then activity and energies are expended in a random manner running in many directions at one time. There is a way to reduce role ambivalence, which may not be available for some of the other role stresses. Since no results come from randomly expended energies and financial commitments, the establishment of a firm priority in which certain activities or roles take precedence over others allows the many obligations and duties to be sorted out by priority, with those most worthy being given more resources than those toward the bottom of the priority list.

While these four role concepts are nowhere near exhaustive of the types of stresses which can be identified in the experiences of Army officers' wives, they do provide some analytical tools whereby some of the most visible pressures and frustrations of the officers' wives may be investigated, identified, and possibly understood more fully as a result of this research.

Research Considerations

This research was conducted over a 3-year period from 1972 to 1975 in two major military installations in west Texas and southern New Mexico. The data were obtained through focused interviews with 50 wives of cadre officers by the junior author while he was serving as a company commander of a small military unit at White Sands Missile Range, New Mexico, and subsequently assigned to the faculty of the U.S. Army Sergeants Major Academy in Ft. Bliss, Texas. Also, a dozen officers were interviewed in depth concerning their perceptions of their wives' role stresses. Through the active participation of the interviewer's wife with the other officers' wives both prior to and following the data-gathering stage, an additional dimension of the informal wife-rank structure was known in both installations. Wives were asked to recall their own personal experiences in the Army which might have involved conflicting loyalties, and they were asked how these were resolved (or not resolved) and with what eventual consequences; such questions were supplemented with hypothetical questions of a standardized nature which distinguished between the problems of role conflict and role strain.[3]

The military ranks of their officer husbands ran the spectrum from 4 Colonels to 6 Lieutenants and 5 Warrant Officers. The enthusiastic support from the commanding officer of both units enabled wives to give full disclosure of their feelings without fear that the information would become public or that there would be reprisals for feelings of discontent. The five wives at one installation who were the only ones not interviewed for various and sundry reasons reflected a broad range of husband military ranks and time in grade that were not unlike the members of the sample population whose cooperation was given.

From information gathered on the interview schedule, role stress indices were developed to determine the varying degrees of role strain and role conflict. Arbitrary numerical values were assigned to the replies to show intensity of these responses. These scores were then analyzed according to the factors of husband's rank, his rank category (i.e., Field Grade, Company Grade, and Warrant), and his time in grade and overall longevity in the military service. These data will now be examined in further detail.

Research Findings

Using the guiding hypothesis that, as the Army officer progresses in his career, the role stresses experienced by his wife would become cumulatively greater and of more intensity, we plotted the patterns of role conflict and of role strain among the sample of Army officers' wives. These are graphically portrayed in Figures 1 and 2.

As indicated in Figure 1, note the wives reported that role conflict situations were fewer during the early years of their husbands' military service, rose to a peak during the "striving years," and declined during the last few years of the officers' career.

When the stresses arising during the various career stages are plotted by rank category of the husband (see Figure 2), the alienation shown by the wife in the earliest years of her husband's career return again during the terminal years, typified by one of the respondent wives of our research population:

> Why should I worry about coffees, teas, formal functions, or making impressions on anyone? My husband is getting out as soon as his obligation is over. It really makes no difference to us, and he has told me not to be concerned about the so-called consequences.

The rationale expressed among the younger officers of "putting in their time" is interpreted by their wives as a "passport for

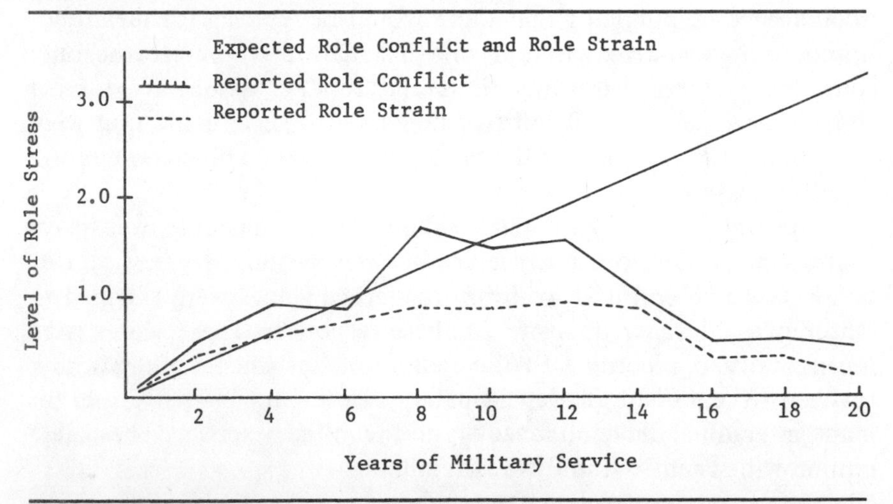

Figure 1: ANTICIPATED VERSUS REPORTED ROLE CONFLICT AND ROLE STRAIN AMONG OFFICERS' WIVES BY YEARS OF MILITARY SERVICE

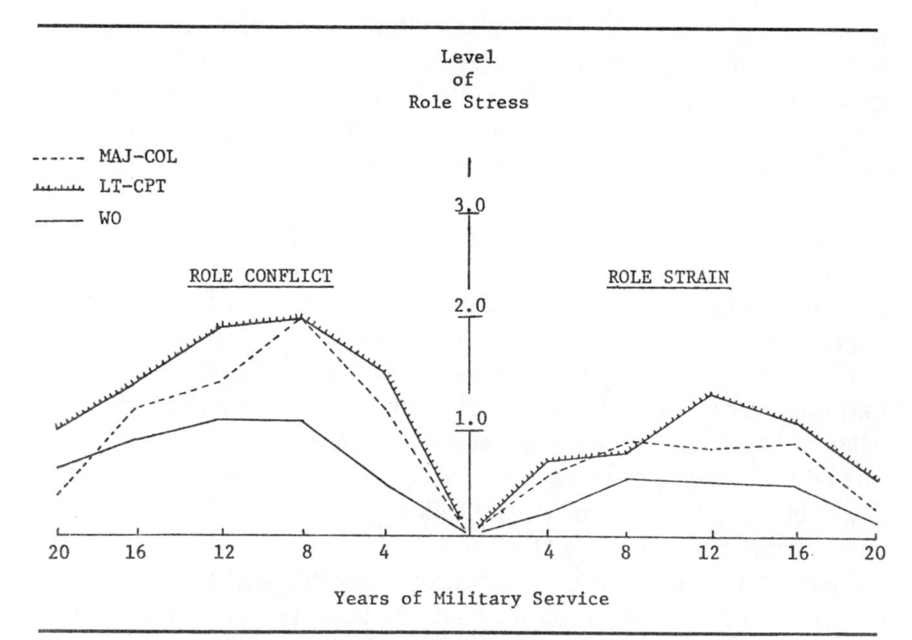

Figure 2: OFFICER WIFE STRESSES REPORTED WITHIN HUSBAND RANK CATEGORIES BY YEARS OF MILITARY SERVICE

noninvolvement." The officer's wife initially feels little or no allegiance to the organization. However, after the initial 4 to 5 years, it becomes evident that the husband has decided to make the service his career. It is at this stage of her husband's career that his wife begins to reconceptualize her role in terms of "helping her husband to succeed by standing behind him as part of the team," and her role stresses begin to mount. The wife is now as committed and inextricably involved as her husband, and the increasing role stresses begin to take their toll.

Inasmuch as divorce or death of the wife or late marriage may cause a variation between the husband's longevity and that of his wife, the small sample of only four couples whose service years were not identical revealed some interesting findings, although their limited number make these conclusions questionable for generalizing to the Army as a whole. The officer who for some reason finds himself without a wife with a dozen years of military service behind him now marries (or remarries) and at that point in his career has, for all intents and purposes, a "new Army wife." Just like the young lieutenant's wife, she begins her socialization by watching others who are, it is hoped, more senior than she. By coincidence, the commanding officers of both installations where the research was conducted had remarried younger women without prior military experience. In both circumstances the subordinate executive officers' wives were "veterans," having accompanied their husbands through the ranks to their present positions in anticipation of becoming "the CO's wife." Yet in these situations the veteran wives were forced into a role of insubordination to younger, less military-experienced women, resulting in exaggerated stresses and fears that their husbands might retire without these veteran wives' having had the opportunity to be the senior wife among wives.

Not only are the initial Army years crucial to the officer's wife, but those last few years prior to retirement are ambivalent ones wherein key decisions are made concerning retirement after 20 years or extending beyond. When the officer decides to "wrap it up after 20," his wife, for the most part, follows his lead.[4] However, the more common pattern was that the wife decided to retire, and this influenced the decision of her officer husband. Within the formal gossip channels of who is doing what to whom, the wife can gather a fairly accurate picture of her husband's terminal rank or potential for "making it" as a general officer. Although the husband may have some inklings of his limitations, true to his male ego he will not give up or concede that his career is over unless he can rationalize his decision. The wife's ultimatum that she will no longer actively support his military aspirations may well prove a source of relief to the officer who quits "because of his family's needs."

As the officer's wife moves closer to retired and civilian statuses, her alienation from the military rituals increases and her role stresses predictably decline. Commonly expressed statements reveal a sense of relief from having to "play the game" any more, as this respondent typifies:

> We have decided to retire, so I need not impress anyone any more. My husband cannot possibly make any more rank, so the whole atmosphere in the house is much more relaxed, and the organizational demands upon us are now secondary.

Implicit within this context is the unusual situation that the colonels and lieutenant colonels in the Army who are most calm and happy are those anticipating retirement. Yet, the Army sees these characteristics as an accurate reflection of the more successful senior officers "on the move." It may well be the reverse, however, in that the general-bound officer and his wife become even more pressured as the stakes in the game mount, with the rewards being rank, status, economic betterment, and power.

That Army officers' wives feel that they, as well as their officer husbands, have "served their 20" in a complementary role to their husbands' careers is shown by this comment of an officer's wife nearing retirement:

> We are settled here now; we are going to retire in this area in 3 more years. It has been decided that if another assignment comes up, he is going by himself. What I do or fail to do now does not affect his career in any way.

When the levels of role conflict among the three broad categories of officer ranks are compared, as in Figure 3, the highest occurs among Company Grade officers' wives with a small decline among Field Grade officers' wives. Wives of Warrant officers, whose ranks are set more by legislative mandate than by husband-wife military involvement, feel less constrained to place military issues, ceremonies, and activities in top priority at the expense of family roles and thus suffer a significantly lower level of role conflict.

What consequences these intense pressures of role conflict spawn may be understood by surveying the means by which officers' wives

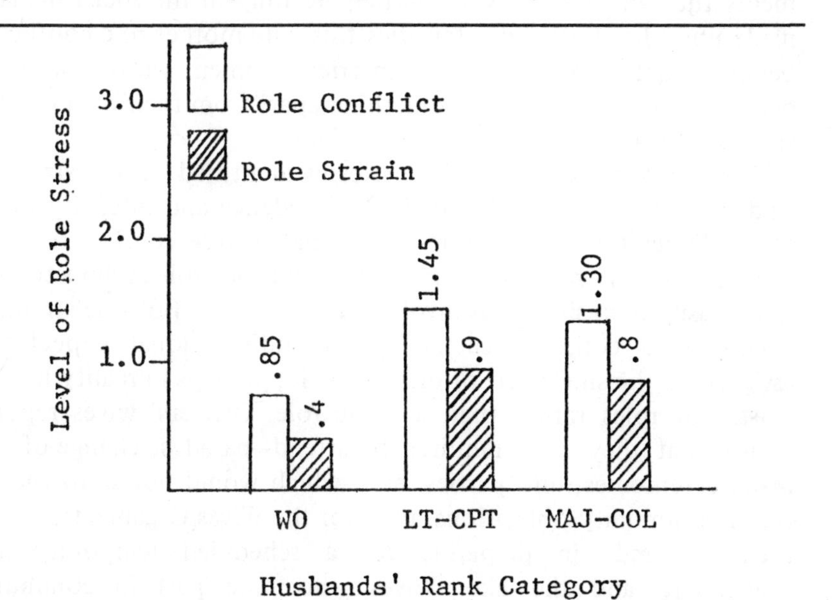

Figure 3: ROLE STRESS OF OFFICERS' WIVES BY HUSBANDS' RANK CATEGORY

resolve them (at the expense of total military commitment). During the early years of her husband's military service, the young officer's wife can use the excuse of pregnancy and child-care duties to excuse herself from participation in all activities which she is expected to attend. Other means employed to legitimate her nonconformity to the informal military demands are those of outside employment, perennial sickness, cyclical, capricious health problems, or attendance at trade schools or institutions of higher learning. Most wives who were interviewed felt that they were not stigmatized for not attending to officer's wife duties if they were working, going to school, or known to be sickness-prone.[5]

The greatest single focus of role conflict reported by officers' wives was that between the mother and wife roles. Even though some of the mechanisms indicated above are employed to alleviate some of the conflict, conflict still persists to some degree. Confronted with a hypothetical situation which would not allow these mechanisms to be employed–a situation requiring attendance at a school play in which a son or daughter was a participant and which occurred at the same time as a command function to which the officer and his wife had been formally invited–more than 80% selected the special event at the child's school, even though such an event does not qualify as a legitimate refusal of a formal military invitation. The rigid requirements that an officer's wife participate fully in the social demands made upon her husband intensifies this wife-mother role conflict far beyond that experienced by married women whose husbands' occupations are considered something apart from their own personal spheres of activity.

Turning our attention to the stresses identified as role strain, we find that, in Figures 1, 2, and 3, the incidence and intensity of role strain difficulties is always less than that of role conflict. The wife respondents also reported being able to manage role strain situations more easily than situations involving role conflict. But role strain still occurs as a consequence of the multidimensional expectations assigned to the single role of officer's wife; perhaps in reality hers is a cluster of roles rather than a single role. Officers' wives reported events that they were required to attend–parades, change-of-command ceremonies, unit parties, etc.–which would not allow them to fulfill community obligations, plan for the Wives Organization, assist their husbands in preparing for a scheduled temporary duty assignment, and the like. Wives who take part in community activities are subject to civilian-controlled scheduling. These often interfere with luncheons sponsored by the Wives Organization, in which they are expected to be not only active participants but also frequently hostesses. These aspects of the Army officer's wife role encompass different obligations, which are perceived as duties of a supportive wife but which are contrary and detrimental to one another. The officer's wife also experiences the normal role strains in which, as a hostess at a gathering, she must show deference to the wives of senior officers while apportioning the largest segment of her time to making the younger wives of her husband's subordinates feel comfortable. The officers in the room carry their ranks on the shoulders of their uniforms, while their wives carry these same ranks as invisible but every present symbols of their informal, *quasi*-official statuses.

A considerable amount of role strain emanates from the incongruence between the formal organizational rank structure of the officers and the informal social ranking of their wives based upon social skills. Figure 4 outlines the wives' status according to the formal position of her husband and then shows the informal "ranking" within the Wives Organization according to such criteria as initiation of projects, decision-making power, pleasant personality, social skills, Army lineage (i.e., the daughter of a high-ranking Army officer currently married to a company grade officer), and other measurements.

In the situation depicted in Figure 4, if Mrs. Colonel at the bottom of the informal schematic is there because she had voluntarily alienated herself from further participation because of the nearness of her husband's retirement, no social repercussions would result. But if this were her ranking because of a cantankerous attitude, crudeness in handling people, lack of civility as a hostess, or similar characteristics, it would cause a major role strain in all the junior officers' wives whose husbands served under her husband. Since the leadership position belongs to the commander's wife in the formal Wives Organization, there should be little question of her assuming it if she chooses to do so. And yet, among the respondent officers' wives in our sample, more than 95% responded that the commander's wife should not always be the leader of the Wives Organization. Clearly, although the formal structure of the Wives Organization may reflect the R.H.I.P. (Rank Has Its Privilege) structure of the officer husbands, it need not determine the informal liaisons among the

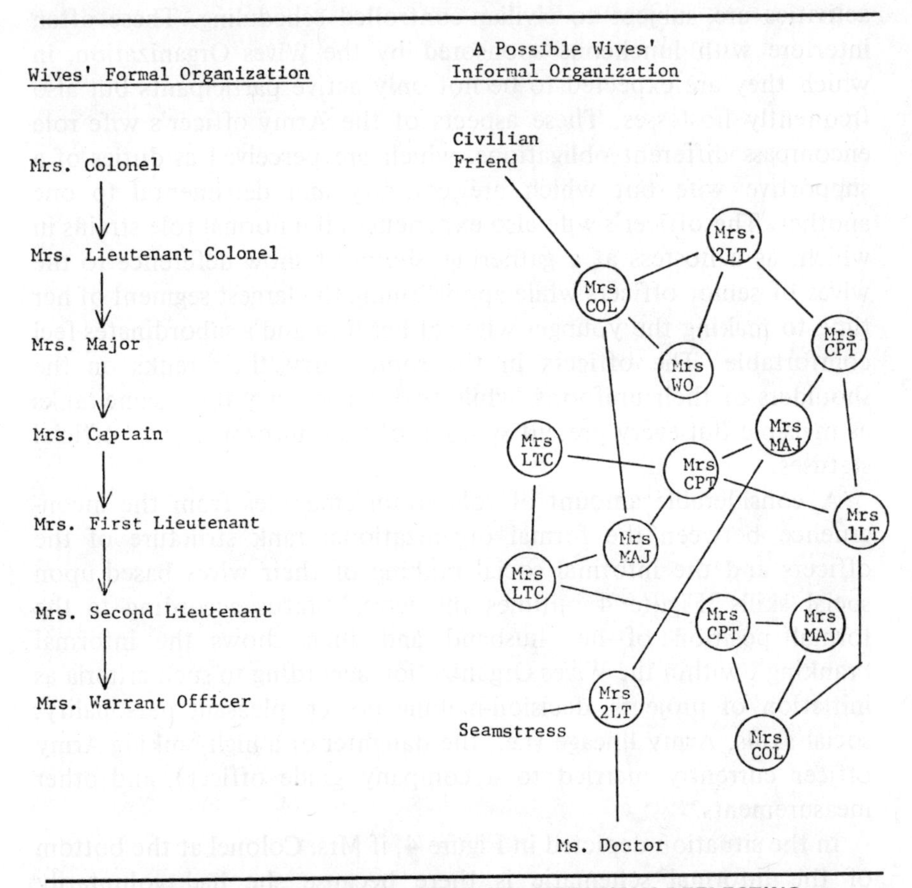

Figure 4: WIVES' FORMAL AND INFORMAL ORGANIZATIONS REFLECTING MILITARY RANK OF HUSBAND

wives themselves in running their own organization—that is, in light of the wives' own comments. Such a liberal attitude creates the milieu for dissonance, exemplified in the following respondent's comments:

> Oftentimes you will find one of the ladies playing up to the colonel's wife when they share something in common like golf or bridge. It can get unbearable for some of the senior wives when they feel the only way to get Mrs. Colonel's attention is through some lieutenant's wife.

It is quite obvious that, if our sample of officers' wives is representative of the Army as a whole, the liberalizing of wife liaisons according to some criterion other than husbands' rank will be eagerly embraced by the younger wives and those somewhat alienated from the present closed system. But it will undermine the traditional expectations of the wife who believes that her reward for helping her husband achieve higher rank is to receive some deference from wives whose husbands are subordinate to him, irrespective of her social graces and outside interests.

In summary, the patterns of role strain reported by officers' wives in our sample followed those established for role conflict, differing only in being less prevalent and less intense. The factor of rank greatly affected the type of role stresses experienced, with the officers still "on the move upward" experiencing more stress while the younger officers still not committed to a military career and the older officers just contemplating retirement being less likely to be caught up in the stresses and strains of conformity. Warrant officers' wives whose husbands are eligible only for in-grade promotions never do approach the stress levels of other officers' wives.

Throughout the length of military service (which is often another facet of the factor of rank) there are different stages at which stresses become more apparent. The beginning of the military career for a wife is primarily a period not of becoming aware of possible promotions but rather of becoming accustomed to the mundane matters of salary budgeting, learning to use the commissary and medical facilities, beginning to participate in some social functions, and the like. But in company with other officers and their wives, certain images of her officer husband and his promotion potential and the rewards accruing therefrom do become more apparent. She rearranges her priorities to give the officer's wife duties more time, effort, and expenditures of all kinds, with the view of increasing her husband's visibility and ensuring his rapid promotion. Once the ceiling of his potential has been reached (whether or not he is aware of it) and nothing more can be gained from personal sacrifices for the advancement of herself and/or her husband, the strains rapidly decrease as she begins to choose friends, interests, and activities (or no activity) to suit herself rather than to suit the ever present specter of her husband's occupational role and her obligations to it.

Our initial hypothesis was that the Army officer's wife would experience a continuously increasing level of role stresses as she played her complementary role to her husband's occupation; this hypothesis was found to be correct only throughout the first half of the officer's career. Thereafter, depending upon perceived opportunities which might result from continued subjugation of personal and family needs to the requirements of an officer's wife, the decision to retire at the end of a 20-year career is accompanied by a rapid decrease in role stresses.

Inasmuch as this study of officers' wives contained no general officers, the trend at the end of the officer's career is skewed heavily downward. For those whose potential for advancement into the general officer category and even continued advancement within it, the years of service–16 to 20–which showed a great deal of alienation among those officers approaching retirement might well be some of the most stress-producing years for the continuously successful officers and their wives. This would be a fruitful area for research in the area of military families. Likewise, it would be valuable to know these patterns among the lower and higher grades of enlisted personnel's wives. Perhaps when the military family becomes a threatening, problematic institution for the military, the military will show greater interest in focusing its research there. Comparing the values and attitudes of our sample of officers' wives with the prescriptive ideals which the Army expects from its officers' spouses, it is not a question of whether or not the Army officer's family life is in difficulty, but rather of when these emergent difficulties will boil over into a visible social problem.

Notes

1. The revised curricula of the service academies now emphasize industrial management leadership training, and the military now employs industrial-type consultants for its organizational problems. These developments might indicate that the military elite has indeed pictured itself as "business managers of warfare." This same idea, held by corporate-successful Secretary of Defense Robert McNamara, proved fallacious during the Vietnam conflict (Halberstam, 1969, chaps. 12-19).

2. Sociologists have coined the term "sick role" to correspond to the social functions of "being sick," whether feigned or real. To claim that one is "sick" releases a person from normal obligations ascribed to her roles–from wishing to avoid a social invitation, engage in social relations, or meet with other officers' wives. This is seen by the military as a legitimate excuse for not having the help of a wife, because it is seen by an officer's wife as a legitimate excuse for circumventing the obligations contingent on her role.

3. For a copy of the interview schedule and details concerning the data collection, consult Claude E. Cabanillas (1975).

4. Subsequent to the field research, in an informal interview on April 17, 1975, a general officer's wife volunteered that her husband had chosen to retire after the 20-year period but that she had persisted in her desire to have him stay on, and now he is a general officer. Just as some officers' wives see a rank ceiling for their husbands, others see the future potential of their husbands' careers even more clearly than the officers themselves do.

5. A research project badly needed in the military medical field is to ascertain how much of the outpatient care, visits, and medicines dispensed to military wives are to legitimize their avoidance of their military-wife duties. A claim of perennial sickness is much less questionable if supported by regular doctor visits, prescription drugs, and visible paraphernalia recognizable as having a medical function.

V.

TRANSITION TO THE CIVILIAN COMMUNITY

POLITICAL PARTICIPATION AND VIETNAM WAR VETERANS: A LONGITUDINAL STUDY

M. Kent Jennings and
Gregory B. Markus

Young adulthood is a time of marked political development. Formal entry is made into the conventional world of politics. For the first time the individual can exercise the franchise and engage in other activities normally associated with adulthood. Patterns of behavior begin to emerge that will characterize the individual for succeeding stages in the life cycle. In the normal course of events, direct and vicarious participation gains momentum during these years, while at the same time the gloss of textbook democracy begins to wear off (Verba and Nie, 1973). It is also a period of substantial change in the individual's life space, including high geographical mobility, educational and occupational ventures, the establishment of a new domicile, and the like. Thus the novitiate period of political participation coincides with new encounters in other spheres of life.

As a consequence of this parallel development, the emerging political character may be strongly affected. One thinks, for example, of the apparent consequences of divergent collegiate experiences, early occupational histories, and residential locations. Although self-selection is a vital element in these effects, the experiential histories also have an importance in their own right. At the very least, one can say that much life-space diversity accompanies the starting-up phase of political participation.

In this paper we will deal with a variation which applies almost exclusively to the male portion of young cohorts, namely, military service. Over two-fifths of all current adult American males have passed through the armed services, and the number of Vietnam-era servicemen alone approached nine million (U.S. Bureau of the Census, 1974). Serving in the military becomes the socialization process most common for postadolescent males. The impact on political behavior of such a widespread phenomenon is clearly in need of assessment.

AUTHORS' NOTE: *The authors wish to acknowledge the financial support of the Ford Foundation, the National Science Foundation, and the U.S. Army Research Institute for the Behavioral and Social Sciences.*

Most of what little is known in a systematic way about the political consequences of military service for the individual lies in the area of attitudes and values (Janowitz, 1971a; Barber, 1972; Jennings and Markus, 1974). In the realm of participation, we know even less. Certainly there is some familiarity with the main-line veterans' organizations and with the newer, more liberal organizations of the Vietnam era. And for the first time in the nation's military history, the Vietnam War brought substantial collective political organization, overt and covert, within the military itself (Oppenheimer, 1971; Moskos, 1973b). That alone was a significant development and surely affected the political makeup of substantial numbers of men. But the evidence is exceedingly thin as to whether and how much individual participation-proneness is affected during and after military service compared with such proneness among civilians.

Not that there is any lack of rhetoric and discourse about the presumed effects of the military experience, especially on postservice behavior. The virtues and vices of serving were debated vigorously during the controversy over universal military training in the late 1940s and more recently in the wave of antimilitarism accompanying (but not solely caused by) the Vietnam debacle. Of relevance for us is the claim of the promilitarists that serving in uniform not only is a citizen duty but also helps produce better citizens, men who will care about and participate in the public life of their democracy. Conveniently, there are just enough national and local examples of soldiers turned activists and leaders to lend face credence to the claim. The counter school of thought paints the military as a rigid, authoritarian, violence-infected institution which dampens or distorts the nascent "civic culture" orientations of incoming recruits. Again, there are sufficient instances of disillusioned, apathetic, and even pathological veterans to win adherents to this view.

Although these contrasting perspectives have more passion than evidence on their respective sides, they will serve as convenient foils for evaluating the results to be presented in this paper. Because of their lack of specificity and theoretical grounding, they will also serve to push us toward a more refined perspective. In this approach we will take into account a number of characteristics surrounding an individual's military history, not the least of which are life cycle factors, sociopolitical contexts, and educational achievements.

Procedures

One of the vexing problems in trying to establish the effect of military service has been the lack of longitudinal data and appropriate control groups. We are fortunate in having at our disposal a set of data that overcome this typical shortcoming. The sample used in the present analysis consists of 611 males first interviewed as part of a larger national sample of high school seniors in 1965 and subsequently recontacted in 1973. Details of the study design and execution are presented elsewhere (Jennings and Niemi, 1973), but it is worth noting here that data were gathered in 1973 from 79% of the males originally interviewed in 1965.[1] A systematic analysis of those who remained in the panel versus those who dropped out (for a great variety of reasons, including death) reveals virtually no differences according to 1965 characteristics. This senior-class male cohort represents only a semicomplete age cohort, because high school dropouts are not included in the study design.

Slightly over half the 611 males in the panel (N = 328) saw active duty during the 8-year period covered by the study.[2] It is possible, then, to observe participation-proneness at each of two points in time. As described below, it will also be possible to make intermediate observations for a variety of specific forms of political participation. Comparisons can be drawn between military and nonmilitary respondents; and, within the military segment, comparisons can be made according to different sorts of military experiences.

The nature of the sample and the nature of the times combine to present us with not only some unique opportunities but also some trying complexities. First, and perhaps most important, the time of entry into the armed forces and the duration of service varied markedly within the cohorts (Table 1). If we consider year of entry as equivalent to age and treat the mean age of the cohort in 1965 as 18, the age of induction can be seen to range from 18 through 23. Similarly, the age at discharge ranges from as young as 19 to as old as 26. When we take into account the general maturation and life history of the 18-26-year-old, the military obviously received and discharged members of this cohort at quite different stages in their personal, social, educational, and political development.

It is equally clear that the historical context was radically different depending upon dates of ingress and egress. Those in uniform during

Table 1. CHARACTERISTICS OF SERVICEMEN FROM THE 1964 TWELFTH-GRADE COHORT

Year of Entry		Year of Discharge		Total Years		Vietnam Duty	
'65	17%	$\leq$'68	21%	$\leq$2	41%	None	45%
'66	38	'69	22	3	2	$<$12 months	16
'67	13	'70	20	4	27	12 months	24
'68	10	'71	15	$\geq$5	9	$\geq$13 months	15
'69	13	'72	9				
$\geq$'70	8	'73	13				
	99%		100%		99%		100%
	(392)[a]		(400)		(392)		(408)

a. Parenthetical entries represent the percentage base. Because of the initial sampling procedures, the data used in the analysis are weighted to a total of 410 for the servicemen and 350 for the civilians. Ns will vary owing to occasional missing data. For a full report on the original design and collection of data see Jennings and Niemi, 1974, Appendix.

the early years of the Vietnam War served in a much more positive and supportive context than did those entering during the later years, when opposition to the war and civil disobedience was rampant and the government was searching desperately for face-saving ways to terminate the strife. By the same token, those returning to civilian status at later stages faced a different reception than did earlier returnees. As the war ground on and servicemen cycled back into civilian life, it became pathetically apparent that it was a war without heroes and without victory parades (Starr, 1973; Helmer, 1974; Polner, 1971; L. Harris and Associates, 1971).

The state of the military itself changed during the 8 years embraced by our observations. By the time the last recruits from our cohort were entering the service, the armed forces had been repeatedly maligned and exposed, desertion and absenteeism rates were approaching the highest levels since World War II, and disrupted authority relationships in Vietnam were widespread. In addition, the members of the armed forces became increasingly better educated, beginning in the 1968-1969 period, as more and more college graduates found themselves in uniform and in Vietnam (Useem, 1973).

Other variations marked the individual experiences of this cohort. One of the most important of these was seeing duty in Vietnam. Nearly half the servicemen were in Vietnam, with the number of months varying considerably (Table 1). Unfortunately, we do not have information on whether they saw actual combat duty. However, the general contention is that whatever political effects military service will have, they will fall more heavily on those more proximate to the war zone.

These variations in timing, sociopolitical contexts, and experiences will be taken into account in the analysis. In the first results to be presented, we have paid particular attention to association between participation-related measures and such factors as duration of service, recency of discharge, time spent in Vietnam, and satisfaction with the military experience. Analysis of these bivariate relationships has been extended via multiple regression to incorporate other relevant predictor variables, including each respondent's 1965 score on the political variable at issue. The multiple regression approach is useful because it allows us to control for exogenous factors which may be systematically related both to the nature of the individual military experience and to the dependent measures of interest. To elaborate, a long line of research has demonstrated the connection between participation-proneness and such factors as educational attainment, familial socioeconomic status, and race. Moreover, civilians and servicemen (and servicemen differentiated by type of duty) are likely to vary systematically in terms of these factors as well. We have employed regression-based adjusted scores on the dependent measures to take into account these kinds of variations.[3]

Our results are divided into two main groupings, distinguished by substance and method. In the first part we take up participative orientations as gauged by the political resources at hand and by the level of spectator involvement in politics. Here the analysis is limited to 1965 and 1973 observations, although we will feel free to interpolate if occasion demands. In the second part we deal with manifest forms of participation and will be able to cast our results in a more dynamic fashion.

Political Resources and Spectator Behavior

It is well known that active political participation varies with a number of individual social-psychological characteristics. To some extent these characteristics are not only preconditions of participation; they are also in a loop which feeds back into participation. That is, certain attitudes, resources, and skills seem to foster participation; and participation, in turn, reinforces the holding of these properties. Our task here is to determine if experience in the military has any net effects on the possession of these attributes. As indicated earlier, we are restricted to before and after measures. We do not know the shape of the distributions during the period of actual military service.

The norm of participation. The principle of civic duty helps propel people into manifest participation quite apart from any rational calculus as to whether or not their behavior will yield tangible benefits. Our indicator of the civic duty norm comes from a question asking the respondents to describe their image of the good citizen.[4] Some of the resulting descriptions had a very heavy political cast, some were non- or apolitical in conventional terms, and still others represented a mixture. Among the political portrayals, some were in the more passive, allegiant mode (e.g., be loyal, obey laws, pay taxes) while others were more active (e.g., vote, campaign, contact officials, be involved). We have taken responses of the latter type to indicate the salience of the participation norm among our subjects. A simple additive index was constructed by scoring each individual according to the number of active political attributes mentioned in their open-ended responses.[5]

At the gross level, whether or not one had been in the service makes no perceptible difference in the degree to which the active mode was stressed. Neither in 1965 nor in 1973 is there more than a hair's difference between servicemen and nonservicemen (Table 2). It

Table 2. POLITICAL RESOURCES AND SPECTATOR BEHAVIOR AMONG VETERANS AND NONVETERANS (Adjusted Means)

	1965			1973		
	Non-veterans	Veterans	Eta	Non-veterans	Veterans	Eta
Political resources:						
Activist emphasis	.92	.95	.02	.72	.74	.01
Political efficacy	6.74	6.49	.05	6.63	6.31	.06
Political knowledge	4.98	4.79	.07	5.07	5.01	.02
Recognition of party differences	.53	.43	.11	.70	.64	.07
General salience of politics	3.34	3.24	.07	3.33	3.40	.05
Usage of media to follow public affairs:						
Radio	3.37	3.15	.06	3.15	3.18	.01
Television	3.73	3.72	.00	4.02	4.21	.08
Newspapers	4.04	4.00	.02	3.98	4.07	.04
Magazines	3.55	3.67	.04	3.46	3.50	.01

is true that the overall frequency of active mode responses declined over the 8-year period.[6] But the drop is virtually the same across both groups. Whatever effect the service may have had appears to have completely disapeared by 1973.

Disguised in this uniformity, however, is a noticeable difference among the servicemen. Those who had served in Vietnam were less likely to stress the active mode in 1973. Moreover, the longer the time spent in Vietnam the more pronounced was the over-time decline, despite relatively equal distributions in 1965. Our final prediction equation for the 1973 results for veterans shows three significant predictors. The partial correlations are all of nearly equal magnitude: 1965 active mode score = .20, respondent's education = .18, and months in Vietnam = –.17.

It is difficult to know exactly what processes are at work in leading to this increasing deemphasis on activism. Perhaps men spending a longer amount of time in the war theater came to feel that loyalty, honor, defense of the country, and the like are exemplary traits because these are qualities which the men were indeed being called upon to exhibit. By virtue of having displayed these traits in greater intensity than other servicemen, the more exposed veterans may project them on to the idealized version of the good citizen.

Another explanation is that these veterans became more upset over the antiwar protests, actions which could easily have been interpreted as acts of disloyalty. Even though disgruntlement and frustration over the war's conduct were widespread among the servicemen, there is at least impressionistic evidence that the servicemen resented being cast as the villains by the antiwar protestors. Such a reaction could lead to more emphasis on the more allegiant, nonactive components of good citizenship. We have no good way of choosing among these alternatives, nor do these two exhaust the likely possibilities. But the evidence is reasonably clear that longer duration in the war theater depressed the salience of the active citizen model.

Political efficacy. Another well-known corollary of participation is sense of political efficacy. Ordinarily, the higher the feelings of efficacy, the greater the level of participation. People are less inclined to take part if they feel that their actions are useless. Our measure of political efficacy is a two-item index utilizing two of the four traditional items on the Michigan political efficacy scale.[7]

Those destined for military service were a bit less efficacious than their peers in 1965, a disadvantage that remains intact in 1973 (Table 2). It is clear that one cannot make an argument that the military enhances the individual's sense of political efficacy. About the best that can be said is that being in the service had little net impact on these men. None of our various within-military distinctions changes this pattern to any significant degree.

Political information. People who know more about political history and the contemporary political scene are more active. Possessing such information seems to give people more confidence; it also reduces the opportunity costs of participation. Here as elsewhere, of course, the process is undoubtedly circular, because more participation tends to lead to more information acquisition.

Our first indicator of political information is an index based on the number of correct answers given to six factual questions asked at both points in time.[8] Nonservicemen were slightly more knowledgeable at each point, with the difference decreasing a bit over time (Table 2). The small relative gain of the servicemen is very probably a normal catching-up that would have occurred in any event. Different military histories are not associated with differential movements within the service group.

Much the same can be said of our other indicator of political information holding. Although the rate of split-ticket voting is increasing and the devotion to one or the other of the two major parties is decreasing, the political parties still stand as the most visible symbols and mechanisms whereby the individual exercises political choice. An accurate understanding of the differences between the parties is a useful though apparently not necessary ingredient of individual choice making and evaluation. In any event, people who grasp these differences are relatively more likely to be active participants in conventional politics. For present purposes we have used a truncated version of the complex series of questions and codes used to develop the "party ideology" scale employed elsewhere. We simple allocate people according to whether they correctly identify the Republican party as the more conservative one.[9]

Incipient servicemen were marginally less likely to recognize the difference in 1965 as well as in 1973 (Table 2). The gap was cut slightly, but even that tiny gain is probably a function of a laggardly segment catching up after a slower start. Differences within the military are negligible.[10] On balance, then, the net effect of military service on this indicator of political resources is nil.

Although most people who take an active part in the political process also observe the game of politics as it is played out through the media and other outlets, the reverse is not necessarily true. Many people are content to watch only from the sidelines; still others do not even engage in these rather passive forms of behavior. Our measures of spectator politics very much hinge on the use that people make of the mass media to keep us with the political world.

Salience of politics. The first and most general indicator is the amount of attention that the respondents claim they pay to public affairs and politics.[11] At neither point in time are the differences between service and nonservice groups very striking, but the relative positioning of the two did change (Table 2). Whereas in 1965 the nonservice group had an edge, in 1973 the servicemen had an edge. Moreover, even the military strata that started out ahead of the general cohort mean in 1965 scored slight gains over time. A particularly interesting civilian-military comparison is that involving the servicemen who were in the military for an extended period of time up to and including the post-1970 period. These men found politics more salient in 1973 than did the civilian cohort (X = 3.5 and 3.3, respectively). The extrapolation here is that long-term service is likely to have a modest salutary effect on the general salience of politics.

Another modest difference crops up among the servicemen. Those spending time in Vietnam expressed more interest in politics in 1973 than those not going to Vietnam. Since there were no premilitary differences in political interest between the two, it seems likely that Vietnam duty served to quicken the political interests of the servicemen. As the results regarding citizenship norms suggest, however, this interest was likely to remain more at the level of passive spectator interest, rather than being translated into a strong participant mode.

Media usage. Turning to more specific illustrations of spectator politics, we examined the use of the various mass media as ways of following public affairs and politics.[12] Use of the print media –newspapers and magazines–is generally recognized as more taxing and as generating more depth than use of the electronic media –television and radio. On the whole there is little difference between the military and nonmilitary portions of the cohort over time (Table 2).[13] Nor are there within-service distinctions of note. During certain phases of service (especially overseas duty), access to some media forms was undoubtedly restricted. These periods of restriction have no long-term effects, however.

We are now in position to summarize our findings regarding the impact of military service on participative orientations. *With the possible exception of general political interest, military service per se makes for scant differences in the 1973 profiles of our 1965 high school senior cohort.* Thus neither of the directional models often found in the rhetoric surrounding the question of military service is upheld. Being in service, at least during the Vietnam period, did not have the salubrious consequences which the advocates of the "military as molder of good citizens" school would have us believe. Being in the service may result in more disciplined and ordered lives later on—as we are often told—but it does little to heighten levels of conventional political awareness and skills beyond what is accomplished through other means also. On the other hand, those who argue—as they did during the Vietnam War—that the service turns people away from politics will not find much support here either. It is simply not true that the veterans emerged as more apathetic and less well fortified with appropriate political resources than did their nonmilitary peers.

A model which seems to fit the data better states that political orientations and resources, as distinct from concrete behaviors, are relatively invulnerable to the impact of transitory, nonvoluntary institutional experiences such as those found in the military. Military service during the Vietnam War would appear to be especially compatible with the outlines of this model. During that war the American soldier was not cut off from the rest of society for very long periods of time. The ever present mass media enabled him to "keep up" if he so desired.[14] Even in Vietnam proper, the rotation plan and the sortielike nature of the war usually meant only intermittent isolation (Moskos, 1970). The antiwar and resistance movements also formed bridges between the military and civilian sectors. In short, the inside and outside worlds intersected frequently. All these factors, coupled with the influence of the premilitary orientations and resource levels already laid down, operated to mitigate the impact that military service might have.

As we pointed out, there were some mild within-service variations around this theme of service and nonservice similarity. The most important of these appeared to be the greater adoption of the nonactive-citizenship norm among those serving in Vietnam, especially among those who were in Vietnam the longest. Another within-service distinction should be introduced at this point, for it is of direct political significance. Recency of discharge is positively related to the 1973 scores on each of the traits taken up in the preceding sections, with the associations being especially pronounced for political efficacy, the two political information indicators, and the use of magazines to keep track of political affairs. At first glance one might hypothesize that those coming out later on were increasingly politicized because of the growing unpopularity of the war and the myriad events surrounding that growth. But two facts alerted us to the weaknesses of that argument. First, the more recent veterans also had higher scores in 1965. Second, and very much related to the first point, the educational attainments of the more recent veterans are dramatically different from those of the earlier veterans. If we divide the time of discharge into the three time periods of pre-1969, 1969-1970, and 1971 and later, the proportions with at least some college training are 30%, 59%, and 77%, respectively. Because education is related to nearly all the political measures, we had every right to suspect a relationship, a suspicion borne out by the regression analysis.

Nevertheless, our exercise has an important bearing on the interpretation of political behavior among veterans and in-service personnel as the war progressed and eventually wound down. Casual observation and some documentation reveal that the vocal expression of opposition to the war, to the system of military justice, to the treatment of returning veterans, and to various other perceived wrongs mounted over time among both veterans and men in service (Barnes, 1972; Yarmolinsky, 1971; Oppenheimer, 1971; Stanton, 1973). This opposition has been variously ascribed to moral and political revulsion, cowardice and disloyalty, contagion from the civilian sector, and even boredom. Some or all of these factors were undoubtedly at work, but our evidence suggests the additional contribution of compositional effects. It has been documented that as the war progressed the proportion of college-educated servicemen grew as a whole and in fact came to be overrepresented in Vietnam (Useem, 1973:91-99). Similarly, in our 1965 cohort the better educated entered later and exited later (as above). These later-arriving men had also been more exposed to the growing discontent in the public and on college campuses and had had more opportunity to engage in a variety of political activities. Since we have observed that the more recent veterans possessed more of the predispositions and skills normally associated with greater involvement and participation, it seems very likely that the upsurge in political expression was greatly abetted by the changing compositional pattern. While it is nearly certain that an upsurge would have occurred anyway —especially in the area of black grievances—we believe that the shift in composition was a catalytic events.

Active Engagement in Politics

In one sense, much of what we have discussed thus far has been a prelude to considering the interplay between military service and the more gladiatorial forms of participation. True, it is not unimportant to ascertain whether the military experience affects attitudes toward participation, the acquisition of political information and skills, and political spectatorship. But the real payoff lies in the determination and explanation of differences in the degree to which civilians, servicemen, and veterans engage in direct forms of participation.

It is likely that the service experience will affect active participatory behavior in a more powerful fashion than it affects the cognitive orientations toward participation already examined. *We hypothesize that, on the whole, military membership increases the costs associated with active political participation.* Illustratively, the inconvenience of absentee voting must discourage a good many servicemen from taking part in normal electoral activity, and overseas duty certainly places limits on opportunities to venture into other modes of political activity as well.

In addition to these relatively objective or physical constraints, military norms—both formal and informal—operate to limit political activity, particularly "controversial" political activity. There is often an uncomfortable lack of fit between pronouncement and reality on the matter of guaranteeing the participatory rights and freedoms of servicemen. For instance, the law says:

> No person may restrict any member of an armed force from communication with a Member of Congress, unless the communication is unlawful or violates a regulation necessary to the security of the United States.

As Peter Barnes (1972:166) points out, however:

> The practicalities of writing to a Congressman are something else again. Soldiers are warned against going outside the chain of command and taking their gripes to a civilian legislator. GI's who insist on exercising this right do so at their own risk. The chances are great that the soldier who "breaks faith" with the army in this fashion will be suitably remembered —and the army has many ways of remembering.

The military fetters the exercise of First Amendment rights of freedom of speech and peaceful assembly as well. Barnes (1972) cites numerous cases to support the argument that "it takes great courage to exercise one's rights as a citizen-soldier in the American army," and Yarmolinsky's conclusions (1971) echo the same theme: "the American military has never felt that broad free speech rights available to civilians could—or should—be granted to servicemen." The latitude of the armed forces to discourage political activity by their members has declined in recent years as a result of court rulings and public reaction, but the vestiges of potent social norms undoubtedly remain.

The fact that we are focusing on a *young* age cohort leads to the point that young adults work up to "normal" levels of participation gradually (Converse, 1969; Verba and Nie, 1973). There are a variety of reasons for the gradual development, including the high geographical mobility of the young, but a major force at work is that of becoming accustomed to participation: participation feeds on participation, and it takes young adults some time to develop the habit. Of course, some are destined to participate more than others, but regardless of whether one is talking about voting or about other more specialized forms of participation, the developmental curve tends to be gradual at least well into the middle years.

This learning model suggests that if impediments are introduced, the rate of learning will be retarded. We have argued that, for servicemen, the patterns of participative development are interrupted or, at the very least, occur under quite different settings than are typical. Hence, the depressing effect of active duty upon participatory behavior may have some residual impact, in the short run anyway, following separation from the service (Converse, 1969).

In studying the effects of service experience on active participation, we will follow our cohort more closely as it moves through time than we were able to in the previous section. This is so because questions in the second wave probing respondents' participatory experiences elicited the approximate dates of such activity. Six modes of political participation are considered here:[15]

(1) Voting in the 1968, 1970, and 1972 elections.

(2) Persuading others to vote in a certain way.

(3) Presenting one's opinions to public officials.

(4) Writing a letter to a newspaper or magazine, giving one's political opinion.

(5) Taking part in a demonstration, sit-in, or protest march.

(6) Working with others to solve community problems.

These activities run the gamut of both traditional and nontraditional measures of political participation. We will examine voting demonstrations and community work separately; the remaining three measures will be combined to yield a factor score tapping traditional participation other than voting.

Because respondents were asked to attach dates to their engagement in these activities, we are able to synchronize the temporal development of participation with individuals' military histories. At a given point a member of the cohort could be either a civilian, a serviceman, or a veteran. It is the ability to link this military status to participation over time that will enable us to observe the consequences for political activeness of being in the military and of being an ex-serviceman.

With a very large sample at our disposal, we could divide the time period under study into a large number of discrete intervals. Unfortunately, the size of the sample and the relatively short historical period covered make it necessary to form just three time periods—1967-1968, 1969-1970, and 1971-1973. These periods embrace two presidential elections, one off-year election, and the height of the antiwar protest movement.

Classifying respondents according to their military status (in service or not in service) at each of the three points enables us to construct seven analytic groupings or subcohorts.[16] In the matrix in Table 3, "+" = in service and "–" = not in service. In addition, 12 men had already become veterans by 1967 and are not represented in the table. Incomplete information on dates of service resulted in dropping another 17 respondents from the distribution.

In the presentation of the results, we will begin with two

Table 3.

Subcohort	1967-1968	1969-1970	1971-1973	Raw N
A	–	–	–	282
B	–	–	+	8
C	–	+	–	33
D	–	+	+	26
E	+	–	–	126
F	+	+	–	80
G	+	+	+	27

aggregates during the first period (1967-1968)–civilians and servicemen. Then as time progresses subcohorts comprising these aggregates will take on unique histories, according to when they enter and/or leave the service. Hence during the second period (1969-1970) there are four aggregates; and by the third period (1971-1973) we shall be observing six discrete subcohorts, each distinguished by a separate history of military experience. (Subcohort B contains too few cases for separate presentation at the third period.)

Reflection on the likely impact of the service experience on participatory behavior leads to a set of predictions about the relative participation rates of the subcohorts at each time interval (Figure 1). Considering 1967-1968 first, we expect those men not in the service (groups A-D) to participate at rates higher than those of contemporary service personnel (groups E-G). This prediction is consistent with the idea that military membership increases the costs of participation. In 1969-1970, men who have seen no service (A and B) should score higher on participation measures than do other respondents, and groups F and G–who have been on active duty since 1967-1968–should exhibit the lowest levels of participation. The 1969-1970 participation rate for group E veternas (+––) should reflect the residual negative impact of their prior military experience, thus placing them somewhat below the "pure" civilians but above the active servicemen. Finally, men in groups C and D should participate at a relatively low rate consistent with their new status as GIs, but they should score somewhat above the long-term servicemen of groups F and G, who have had their participation growth slowed by a longer period of military membership. Looking lastly at participation in 1971-1973, we expect group A (–––) again to score highest, with subcohort G (+++) exhibiting the least participation. Groups E (+––), C (–+–), F (++–) and D (–++) should be arrayed, in that order, between subcohorts A and G, a prediction that reflects

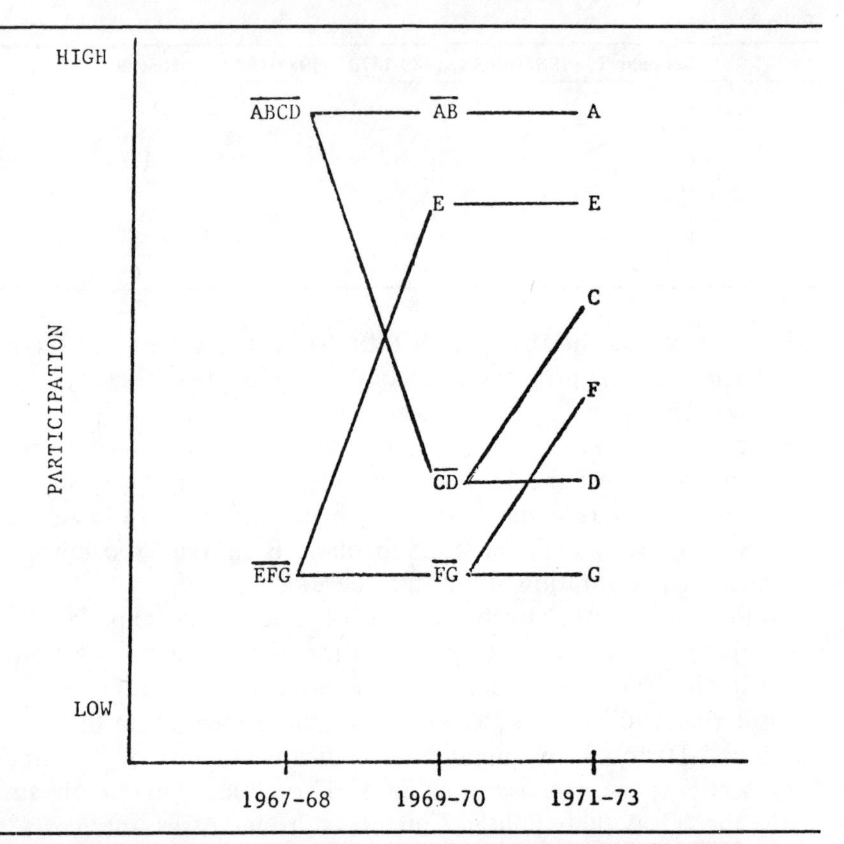

Figure 1: PREDICTED PARTICIPATION RATES OF SUBCOHORTS OVER TIME

both the duration of the service hitch and the span of time since separation.

The accuracy of these predictions depends on the degree to which other sources of variation are held constant. This includes variation across cohorts, participative modes, and time. For example, it is likely that the groups under study differ systematically not only in terms of the temporal location of their service experience but in other ways as well. Men who have spent much of their early adult life in the military are less likely to have college experience, and systematic differences with regard to such demographic factors as socioeconomic background and race are probable as well. To counteract the threats to valid inference presented by such possibilities, participation rates for the cohorts will be "corrected" as in the previous section.

It is also possible that our initial assumptions about the additional costs of participation imposed by the service experience may be more accurate with regard to some forms of participative activity than others. The military-related restraints placed on joining in demonstrations, for instance, are doubtlessly more stringent than those associated with more common forms of political activity. And for some modes of participation, such as community service, it is possible that the armed forces may actually enhance opportunities for action through the creation of veterans' clubs, military base organizations, etc.

Finally, the participation costs associated with the military experience may not always be sufficient to discourage participation, provided that motivation supplied from other sources offsets the added costs. More specifically, subcohort differences in participation rates may be diminished in times when the events of the day give rise to high political stimulation. Or, as the motivation and opportunity to participate increase over the life cycle, a similar reduction of intergroup differences may be observed. Also, the military norms restricting participation have weakened in the recent past, a factor which may be evidenced in our data on 1971-1973 participation.

Having corrected the participation scores for exogenous sources of variation and keeping other caveats in mind, let us turn to the empirical results. Considering voting first (Figure 2), we find that turnout rates in 1968 follow a pattern that is consistent with our expectations, with civilian subcohorts (A-D) voting in greater numbers than their military counterparts, ceteris paribus. In 1970, also, the results are closely in accord with our hypotheses. Civilians exhibit the highest corrected mean turnout, and the long-term servicemen of subcohorts F and G are among the least likely to have voted. Category E (+––) displays the postservice decrement in participation that was hypothesized. Finally, the largest drop in adjusted turnout rate occurs for those former civilians who entered the military in the 1969-1970 period (subcohorts C and D).

In 1972, however, intergroup variation in adjusted turnout rate is quite small. It is true that the "pure" service category evinces the lowest proportion voting, but the difference is marginal. In attempting to explain this result, we argue that the high stimulation of a presidential election coupled with the advancing maturity of the cohort overwhelmed any variation in turnout that would be attributable to the service experience.

We turn next to the three-component participation factor score

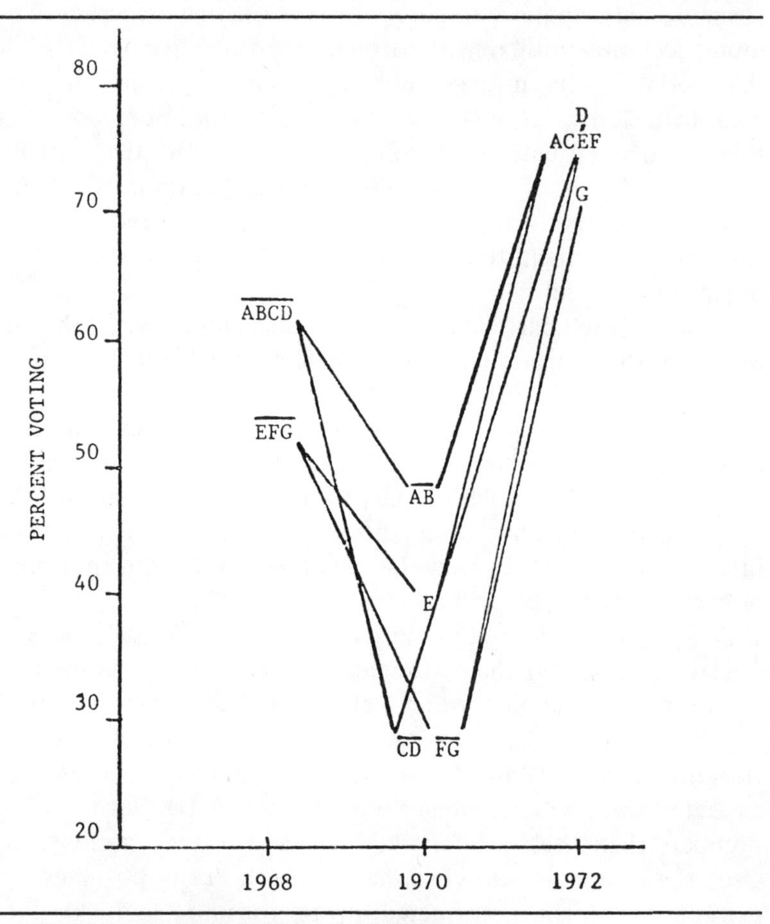

Figure 2: ADJUSTED VOTING RATES OF SUBCOHORTS

(Figure 3).[17] As with voting, it is apparent that the 1967-1968 civilian males score somewhat above their military counterparts. In 1969-1970 the "pure" nonservice subcohorts (A and B) are again the most active participants, and the "pure" military aggregate is least active. Note, too, that although the veterans of category E do not display any decrement in 1969-1970 participation owing to their service experience, the new recruits of groups C and D lie between the civilians and the more senior servicemen as hypothesized. In contrast to these generally regular results, in 1971-1973 the subcohorts are not arrayed in any consistent pattern. As argued above, a combination of life cycle and short-term forces (perhaps coupled with a weakening of antiparticipation military norms)

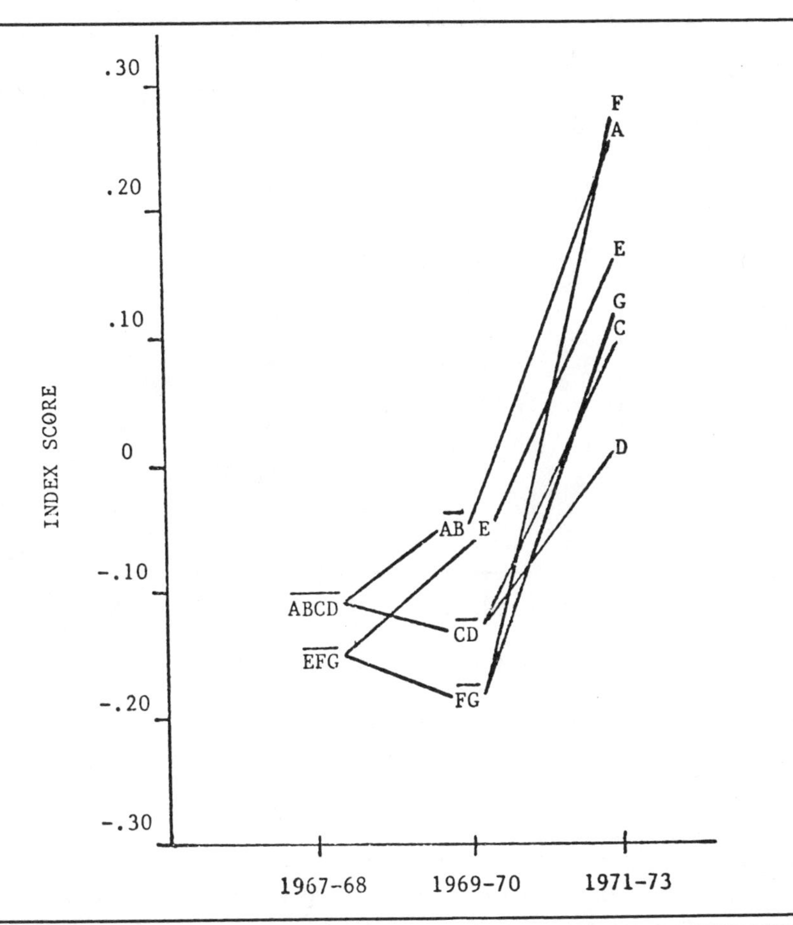

Figure 3: ADJUSTED PARTICIPATION INDEX SCORES OF SUBCOHORTS OVER TIME

probably account for this apparent anomaly. In any event, it is clear that, under certain conditions, military service need not *necessarily* depress political activism.

A prima facie candidate for a clear-cut display of the service-participation effect is individual participation in demonstrations. Many military officials take a dim view of public protest in uniform, the ambiguity of the law notwithstanding. And even participation in demonstrations by GIs in mufti has brought quick military reaction (Barnes, 1972; Yarmolinsky, 1971). Indeed, Figure 4 shows that the adjusted proportion of servicemen taking part in protest activity in 1967-1968 was markedly lower than that of civilians in the same period. In 1969-1970 as well, men with no service experience (groups

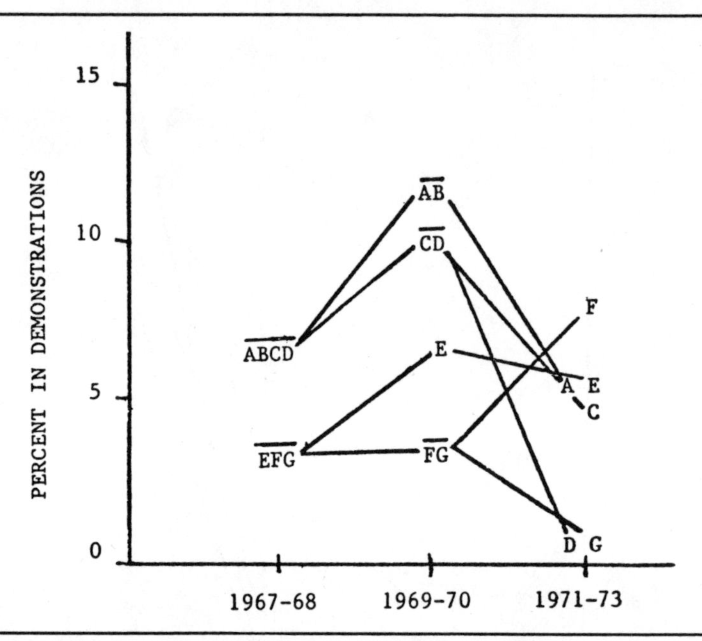

Figure 4: ADJUSTED DEMONSTRATION RATES OF SUBCOHORTS

A and B) were much more likely to demonstrate than were long-term servicemen of the cohort. The young veterans of category E were also not prone to take part in such behavior.

What *is* unexpected, however, is the relatively high proportion of new recruits (groups C and D) who were involved in some form of protest activity in 1969-1970. We may speculate that a fair number of these recruits resented being drafted (or coerced into enlistment by threat of the draft) into a growingly unpopular war and that their bitterness was manifested in demonstrations, despite the threat of military sanctions. The fact that a number of men in this subcohort had previously engaged in public protest (as attested by their 1968-1969 scores) could only have served to facilitate their continued participation in such activity.

Our argument finds support in the fact that, looking only at the 1969-1970 behavior of subcohort C (–+–), we find that *this group was the most likely of all subcohorts to engage in protests* (particularly antiwar protests).[18] One should recall that this group consists of short-term recruits who returned to civilian life 2 to 3 years hence. In contrast, the subcohort D (–++) adjusted participation rate of 4.5% in 1969-1970 is identical to that of the longer-term GIs. Thus we have located in subcohort C a sample of men who were inducted into a war which many of them opposed, who displayed their opposition by public protest, and who subsequently returned to civilian life at the first opportunity. Once again we are faced with an illustration of the thesis that, although the military experience may increase the costs of participation, these costs are insufficient to deter highly motivated servicemen.

In 1971-1973, protest activity decreased among all subcohorts, and respondents who remained in the service during this period were particularly unlikely to engage in demonstrations. Of the remaining groups, we find that it is not the "pure" civilian aggregate who possess the highest demonstration rate but the recent returnees of subcohort F; and if only *antiwar* demonstrations are considered (not shown), the results are the same. It appears, then, that even some of these veterans had finally come to oppose the Vietnam War, resulting in their protest activity's "peaking" somewhat later than for other men in the sample.

The final measure of political activism that we examine is of a different sort. It deals with participation in community or neighborhood affairs. Earlier in this section, we suggested that the military experience may actually enhance community participation by integrating its members into a common social network. Figure 5 indicates that this may indeed be the case. Beginning with a trivial difference between civilian and military aggregates in 1967-1968, we find that the service subcohorts (F and G) move to a relatively high level of community participation in 1969-1970. The differential effects of the service experience on community participation versus other political behaviors is underscored more dramatically in 1971-1973. At this time, the veterans of long service (subcohort F) exhibit the highest participation rate, and the long-term servicemen of group G are next in line. Although the servicemen of group D trail the pack, we expect that their community involvement will increase with their service tenure. We suggest, then, that the service experience—by establishing a common social environment for previously diverse individuals and by enabling a continuity of social experience even across differing geographical locations—has a salutary effect on community involvement.

In examining the longitudinal interrelationships of military service and active participatory behavior, we focused on three types of variance. First, we studied differences in participation rates across

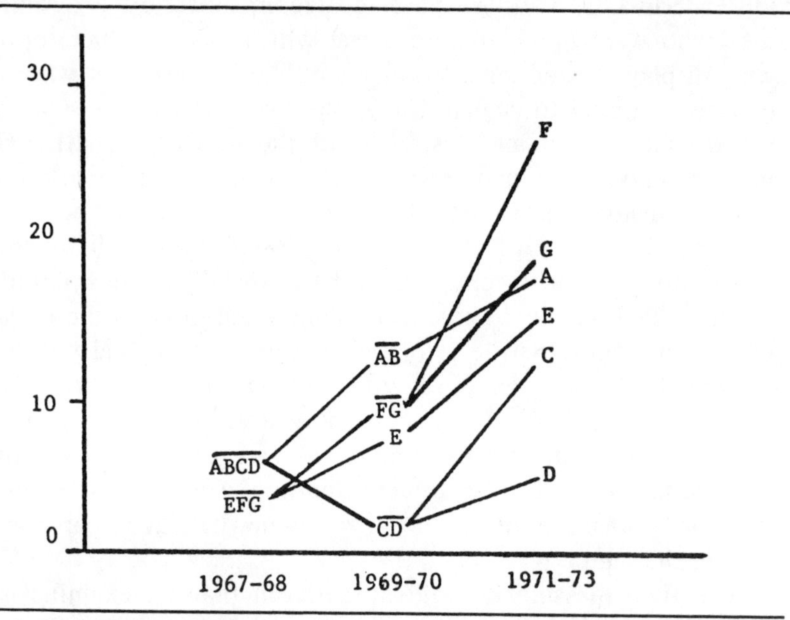

Figure 5: ADJUSTED COMMUNITY PARTICIPATION RATES OF SUBCOHORTS OVER TIME

subcohorts with differing histories of military experience. We found that, *generally speaking,* there are additional costs of participation that are imposed on the individual as a result of his membership in the military—costs that may stem from the physical constraints of military life as well as costs whose origins are in the norms which are part of the military social structure. This led to a concern with how other factors might modify this basic pattern. Thus we looked also at variation in the *mode* of participation. It was noted that at least one form of participative behavior, community service, did not follow the general rule and that, in fact, a tendency of the opposite kind was found. Finally, we were sensitive to variation across time. In a number of instances, the political nature of the times and the growing maturity of the cohort provided a motivation to act that overrode any costs of participation linked to military membership. Elaborations such as these may serve to make the simple hypothesis more complex, but at the same time they yield greater insight into the process by which an individual comes to engage in politics.

Conclusion

The variability of the participation-service nexus is an appropriate note upon which to close our discussion. We have been concerned with the effects of a particular institution on its members when these members are at an important developmental stage of the political life cycle. The question confronting us was whether being in the military, and specified conditions thereof, would lead to different orientations to and engagement in political participation. We were fortunate in having at our disposal longitudinal data from a cohort heavily hit by the increasing personnel demands of the armed services as the Vietnam War escalated. Given the basic division of the cohort sample into nonservice and service components, plus the dramatic differences in the timing and experiential histories attached to the latter, we were able to explore in some detail the consequences of this particular institutional experience on the emerging character of the cohort's participation-proneness.

Serious as well as popular generalizations about the political impact of serving in the armed forces often spring from specific examples and deviant cases. Our inquiry has revealed that gross extrapolations are improper. Even the generally (though not universally) null relationships uncovered in the domains of political resources and political spectatorship were important as comparison points for the more vibrant relationships in the area of active participation. We also found, in treating those domains, one of the keys to understanding why militancy grew among servicemen and veterans over time. And, as we have just observed, there are marked variations when considering active participation. Of central importance in those variations were several dimensions of time.

Given the foregoing points and the still unfolding nature of our cohort's political makeup, we are loath to speak of "the" consequences of military service for individual political participation. Still, it is clear that the starting phase of participation is affected by in-service status. What is less clear—and what is of considerable public interest—is the impact on postservice participation. In the domains of resources and spectatorship the net impact—within our time frame—appears to be variable, but generally negligible. In the domain of active engagement, the results are more complex and, in a very real sense, not yet final because regularized patterns of active involvement—as opposed to more passive forms—await the passing of time and the participation opportunities that come with that passage. *On balance,* however, one could say that the postservice consequences are at least not deleterious. Indeed, some subcohorts participate at higher than expected levels after discharge. But it is precisely variations of this sort which warn us against easy generalizations.

Notes

1. Of that total, four-fifths were personal reinterviews and the remainder were lengthy mail-back questionnaires.

2. An additional 63 respondents were in the National Guard or in the Reserves during this period, but they have been dropped because of their predominantly non-active-duty status.

3. The correction procedure involves regressing the dependent variable of interest on three control variables plus a set of binary variables identifying the service-related category of each respondent. The three control variables are a binary (white/nonwhite) race variable, a binary education variable (no college/some college), and the Duncan SES scale of father's occupational status (range = 0-97). The adjusted dependent variable score, $\tilde{y}i$, is calculated as:

$$\tilde{y}_i = A_g + b_1 \overline{SES} + b_2 \overline{RACE} + b_3 \overline{COLLEGE} + e_1$$

where bars over variables denote sample means, A_g is the intercept for the service experience group of which individual *i* is a member, and e_i is the residual for individual i derived from the original regression equation.

4. The question read: "People have different ideas about what being a good citizen means. We're interested in what you think. Tell me how you would describe a good citizen in this country–that is, what things about a person are most important in showing that one is a good citizen?"

5. We experimented with a number of other indices, some standardizing for the total number of answers, others using ratios. The intercorrelations among the various indices were all above .9, so we have retained the simpler technique for present purposes. The range of the variable is 0-4.

6. This decline is not inconsistent with the developmental model of growing participation during these years. Rather, it represents a normal movement away from the impossibly high standards stressed in public rhetoric and in the classroom. See Jennings and Niemi (1974: chap. 10).

7. The two items were: (1) "Sometimes politics and government seem so complicated that a person like me can't really understand what's going on"; (2) "Voting is the only way that people like me can have any say about how the government runs things." The index ranges from a low of 2 to a high of 10.

8. These questions were: (1) "About how many years does a U.S. Senator serve?" (2) "Marshall Tito is a leader of what country?" (3) "Do you happen to know how many members there are on the United States Supreme Court?" (4) "Who is the governor of this state now?" (5) "During World War II, which nation had a great many concentration camps for Jews?" (6) "Do you happen to recall whether President Franklin Roosevelt was a Republican or a Democrat? (which?)"

9. The wording sequence ran as follows: "Would you say that either one of the parties is more conservative or more liberal than the other?" [IF YES] "Which party is more

conservative?" [IF NO] "Do you think that people generally consider the Democrats or the Republicans more conservative, or wouldn't you want to guess about that?"

10. One minor but intriguing difference is that in 1965 the men eventually bound for Vietnam were a shade more likely to designate the Republicans as more conservative; by 1973 they placed the Democrats there by the same margin. Serving in Vietnam created a slight blurring effect, perhaps because of being away from the immediate political scene or because of the similarities in the Johnsonian and Nixonian war postures.

11. The question ran as follows: "Some people seem to think about what's going on in government and public affairs most of the time, whether there's an election going on or not. Others aren't that interested. Would you say you follow what's going on in government most of the time, some of the time, only now and then, or hardly at all?" The variable score ranges from 1 to 4.

12. Attentiveness to each of the four media was coded as follows: (1) Never, (2) A few times a year, (3) Three or four times a month, (4) Two or three times a week, (5) Almost daily.

13. The one marginal exception is the slightly greater gain on television usage among servicemen, even though in 1965 the two groups had virtually identical usage rates. Significantly, this greater increment occurred for one of the "easier" media forms.

14. Janowitz (1971a:200) makes this point for the pre-Vietnam buildup period as well. So far as we know, however, the actual communication world of servicemen has not been studied, and in terms of wartime service it seems likely that the Vietnam servicemen were less isolated from the communications net than were previous wartime servicemen.

15. The exact wording of the questions was:

"1. (a) In talking to people about the presidential election last year between Nixon and McGovern we found that a lot of people weren't able to vote because they weren't registered or they were sick or they just didn't have time. How about you, did you vote or did something keep you from voting? (b) How about the election for Congress in 1970–did you vote for a candidate for Congress that year? (c) Now in 1968 you remember that Mr. Nixon ran on the Republican ticket against Mr. Humphrey for the Democrats and Mr. Wallace on an independent ticket. Do you remember for sure whether you voted in that election?

"2. I have a list of some of the things that people do to help make an election come out the way they want it to. I wonder if you could tell me whether you have done any of these things during *any* kind of public election since about 1965? By that I mean elections for public office or votes on issues, propositions, referenda, and so on. First, did you ever talk to any people and try to show them why they should vote one way or the other? [IF YES] When was that?

"3. Aside from activities during election campaigns, there are othr ways people can become involved in politics. For example, have you ever written a letter or talked to any public officials, giving them your opinion about something? [IF YES] When was that and what was that about?

"4. Have you ever written a letter to the editor of a newspaper or magazine giving any political opinions? [IF YES] When was that and what was it about?

"5. Have you ever taken part in a demonstration, protest march, or sit-in? [IF YES] When was that and what were the circumstances?

"6. Have you ever worked with others to solve some community problems? [IF YES] When was that and what were the circumstances."

16. In classifying respondents, we employed the rule that, if a respondent was in the service for any part of a year, he was coded as being "in service" for that year.

17. The factor score equation was based on an analysis of the pooled participation data for the three time spans. The equation for the participation index, PI, is:

$$PI_{it} = -.345 + .549\ TALK_{it} + 1.39\ CONTACT_{it} + 1.56\ WRITE_{it}$$

where the three independent variables are binary and denote whether or not individual *i* engaged in the given activity at time *t*.

18. The subgroup C corrected rates are: 16.1% for all protests and 14.3% for antiwar protests.

THE IMPACT OF MILITARY SERVICE ON TRUST IN GOVERNMENT, INTERNATIONAL ATTITUDES, AND SOCIAL STATUS

David R. Segal and
Mady Wechsler Segal

Military Experience and Social Attitudes

Two major controversies have developed about the social-psychological effects of military experience. The first consists of inconclusive results regarding authoritarianism and conservatism. While authoritarian personalities have been shown to have preferences for the military (Roghmann, 1966:254 ff.) and military men have been shown to be relatively conservative (Abrahamsson, 1970), it is not clear whether this effect is wholly due to self-selection or whether being in the military contributes to conservatism and authoritarianism. Research has been conducted showing no change (French and Ernest, 1955), increase (Christie, 1952), and decrease (D. Campbell and McCormack, 1957) in authoritarianism as a function of military service. A replication of Campbell and McCormack followed a panel of recruits in the West German Army from 1966 to 1968 and likewise found a decrease in authoritarianism (Roghmann and Sodeur, 1972). While there has been debate on the interpretation of these findings (Stinchcombe, 1973; Roghmann and Sodeur, 1973) the direction of change is not questioned, and the current state of the art presents no strong basis for asserting that military service changes attitudes in a conservative direction.

The second controversy revolves around a more timebound concern with the effects of the Vietnam War. On one side is a series of studies suggesting that veterans of this conflict returned hostile toward the government, disrespectful of human life, alienated, and disoriented (see, e.g., Lifton, 1973a; Helmer, 1974). However, in contrast to this view, we find research reporting that middle-aged civilian women are more favorably disposed to violence than Vietnam-era enlisted personnel (Brady and Rappoport, 1973), that Vietnam veterans are on the whole less cynical and more cosmopolitan than their nonveteran peers (Jennings and Markus, 1974), and that in general, veterans of the Vietnam era are more similar to than different from nonveterans in their age cohorts (Bachman and Jennings, 1975).

Military Experience and Civilian Income

A third controversy over the effect of military experience is concerned with the effects of service on civilian income. Economists have generally regarded the individual who has served in the military and then returned to the civilian labor force as having been penalized for his service by experiencing a loss to lifetime earnings that may run as high as $9,000 (e.g., J. Miller and Tollison, 1971). Browning et al. (1973), on the other hand, have found that among men in the states of Arizona, California, Colorado, New Mexico, and Texas, blacks and Chicanos experienced income increments as a function of military service, although Anglos did suffer an income penalty. Between these two positions, Mason (1970), Kassing (1970), and Cutright (1974) have all indicated that there are no significant differences in earnings between veterans and nonveterans when other factors, such as age, are taken into account.

Much of the literature presenting veteran and nonveteran comparisons contains only limited information on military service. Where extensive information on such factors as the duration, recency, intensity, and affective element of the military experience is available, as in the Jennings and Markus and the Roghmann and Sodeur panel studies, only a thin slice of military history is dealt with. That is, all the respondents in these studies served in the military in the same era. The present study brings to bear on these problems a set of extensive data on military experience from a cross-section sample of the Detroit population. While we lose the strength of the before-after design built into panel analysis, we gain a broader time perspective unavailable from samples followed for a decade or less. The major social-psychological variables to be considered are trust in government and isolationism-interventionism. In addition, we shall consider the effect of military experience on education, occupation, and income.

Data Base

During spring, summer, and fall 1973, a survey was conducted in the Detroit area by the Detroit Area Study of the University of Michigan. Between May and October, 576 interviews were conducted in an area including about 85% of the population of the Detroit Standard Metropolitan Statistical Area covering portions of Wayne, Oakland, and Macomb counties. The respondents were persons 18 years of age or over, residing in housing units. A multistage-stratified-clustered sampling design was used to sample 109 blocks (sampling fraction = 1/10,700). Approximately 8 housing units per block were selected for interviews.

Of 849 households included in the sample, 32 were vacant and 9 were errors in the household listing from which the sample was drawn. Of the remaining 808 houses constituting the actual sample, our interviewers encountered 176 refusals, and in 56 cases interviews were not conducted because no one was found at home after repeated call-backs, because the appropriate respondent was absent from the home or for other reasons. Thus, our response rate was 576/808 = 71.3%.

The sample included a 20% subsample of veterns (N = 117). The veterans were predominantly male (97%), were of significantly higher socioeconomic status, but had the same racial composition as the nonveteran subsample (D. Segal, 1975a). Table 1 shows the period of entry into military service of the veterans in our sample. While mortality has clearly taken its toll of pre-World War II cohorts, there is a wide range of military experience represented in our sample.

Our veterans were predominantly from the Army (64.7%), with the other services being less represented (Navy = 12.9%; Air Force =

Table 1. PERIOD OF ENTRY INTO MILITARY SERVICE*

	N	%
Pre-World War I (1900-1916)	–	–
World War I (1917-1918)	2	1.8
Interwar (1919-1940)	1	.9
World War II (1941-1945)	40	35.4
Interwar (1946-1949)	5	4.5
Korea (1950-1953)	10	8.9
Interwar (1954-1964)	39	34.5
Vietnam (1965-1974)	16	14.2
Total	113	100.2

*Four veterans did not indicate year of entry.

7.8%; Marines = 6.9%). A majority (55.2%) of our veterans were conscripts. More than half (56.5%) had been in combat arms (e.g., infantrymen, pilots, unrestricted line naval officers), 18.3% in combat support (e.g., military police, engineers, military intelligence personnel), and 25.2% in combat service support (e.g., financial and medical personnel, servicewomen, chaplains). The predominance of combat specialties in our sample is reflected in the fact that almost half (47.7%) of the veterans reported that, after discharge, they made no use of job skills acquired in service.

Two-thirds of our veterans had served overseas. More than half (54.8%) had never been in combat zones. Another 15% had been in combat zones, but not under fire; 30% had been under fire.

Military Service and Attitudes

Our respondents were asked their opinions on a wide variety of issues relevant to military institutions, the Vietnam War, international relations, and the U.S. government (see D. Segal, 1975c). The major dependent variable used by Bachman and Jennings (1975)—trust in government—was central to the set of attitude data that we collected. We hypothesized that the differential effects of the Vietnam War on veterans and nonveterans would be no different from the effects of other wars. Thus, since Bachman and Jennings found no major differences between Vietnam veterans and their nonveteran peers, our expectation was that we would find few if any differences between the veterans and nonveterans in our Detroit sample.

We compared the male veterans (N = 113) with the male nonveterans (N = 124) on 23 predominantly Likert-type attitude items covering trust in government and isolationism-interventionism. There were significant differences between the two groups ($p < .05$) with regard to only two of these. Nonveterans were more likely to approve of replacing the draft with an all-volunteer Army (87%) than were veterans (78%), although overwhelming majorities of both groups supported the conversion. Nonveterans were also more likely (35%) than were veterans (20%) to agree with the statement that "Stationing some of our American troops and planes in Europe is of no help to the United States." Clear majorities of both groups disagreed with the statement.

Were we to push our data very hard, we might argue on the basis of these findings that veterans are somewhat more traditional than nonveterans in their view of the military, in terms of both greater support for conscription and greater support for a U.S. military presence in Europe; and some people might regard this as conservatism. These views are minority positions even among the veterans, however, and finding 2 comparisons out of 23 significant at $p < .05$ is very close to chance occurrence. The most striking finding, as in the Bachman and Jennings data, is the absence of differences between veterans and nonveterans.

Components of Military Experience

Within the veteran subsample, relationships were sought between components of the military experience and attitudes relating to (1) governmental support and (2) isolationism-interventionism (including the appropriate role of the military in foreign affairs). These two sets of attitudes have been shown to be basic in the ogranization of national security opinions (Modigliani, 1972; D. Segal, 1975c). On the basis of Modigliani's factor analyses of other survey data, we had anticipated—and the factor analysis revealed—two factors reflecting attitudes toward the government and the way it was doing its job, as well as attitudes about America's role in world affairs. Scores for each of these factors were computed for each of the veterans in our sample by standardizing the variables with high loadings on the factors and by multiplying the standardized measures by the factor scores for those variables derived from the total data set.

The attitudes in the government support score centered on whether Americans had been asked to make too many sacrifices to support the defense program, whether officials in Washington were satisfactorily handling foreign affairs, whether the government is doing all it can to avoid another war, whether the government is doing all it should to prevent the spread of communism, and whether the government acted as quickly as it could to arrange the release of our POWs from Vietnam. The items reflecting isolationism-interventionism centered on whether it is more important to stop communism than to stay out of war, whether women should bear arms, whether the U.S. should play a major role in maintaining world order, and whether it was a mistake to send U.S. troops to Vietnam.

The relationships between these factors and a wide range of components of military experience were analyzed. The components included service, branch, primary job in the armed forces, military occupational specialty, primary assignment, nature of military accession, reasons for enlisting (for recruits), feelings about being drafted (for conscripts), period of service, rank at discharge, highest rank attained, transferability of military skills to civilian job, combat experience, service overseas, and postservice membership in veterans' organizations.

While most of this analysis is exploratory and descriptive, we do have several hypotheses on which our data may be brought to bear. In their analysis of Vietnam veterans, Bachman and Jennings (1975) found that (1) mode of entry into the military had little or no effect on trust in government; (2) combat experience had no effect on trust in government; (3) differences existed among branches of the military service, with Air Force veterans experiencing the most severe drop in government trust and the Navy least. Although our data are cross-sectional rather than longitudinal, and our measure of government support differs from theirs, these conclusions can be taken as hypotheses to be tested with our data. In addition, theories of occupational socialization and of cognitive dissonance would predict that (4) men who had served in combat arms would take a more interventionist view of military affairs than those who served in other branches; (5) volunteers would be more interventionist than conscripts; (6) men who had served in combat would differ in their views of the military role from those who had not been under fire.

Results

Our findings on these relationships are summarized in Table 2. We find that service is related to government support, with the rank order of the services reflecting the Bachman and Jennings results. Government support scores ranged from –5 to +4, and the means for the services were: Air Force, –.112; Army, –.069; Navy, +.420. Like Bachman and Jennings, we found no effect owing to combat experience and no difference between conscripts and recruits. However, persons whose primary assignments had been in tactical or medical units had significantly lower government support scores than those who had served in other types of units. It is perhaps not

Table 2. SUMMARY TABLE: COMPONENTS OF MILITARY EXPERIENCE AND ATTITUDES

Experience	Government Support	Isolationism-Interventionism
	tau	*tau*
Service	.0959	n.s.
Branch (Army only)	n.s.	n.s.
Combat/combat support/combat service support	n.s.	n.s.
Primary assignment	.0873	n.s.
How entered military	n.s.	.0276
Reason for enlisting (recruits only)	.2088	n.s.
Reaction to draft (conscripts only)	n.s.	n.s.
Reduction in rank	.0131	n.s.
Service overseas	n.s.	n.s.
Member of veterans' organization	n.s.	n.s.
	gamma	*gamma*
Civilian counterpart occupation	.2171	.1960
Period of service	n.s.	n.s.
Rank when discharged	n.s.	n.s.
Highest rank achieved (if higher than at discharge)	n.s.	n.s.
Degree of skill utilization in civilian job	.2618	n.s.
Degree of risk experienced in combat	n.s.	n.s.

All associations shown are significant at $p < .05$ using chi-square for nominal variables and Kendall's tau for ordinal variables. Goodman-Kruskal tau is shown as the measure of association for nominal independent variables, Goodman's gamma for ordinal independent variables. Both of these measures indicate proportion reduction in error.

surprising that these patterns replicate most of the Bachman and Jennings findings on Vietnam veterans, since we also find no effect owing to period of service: Vietnam veterans in our sample were not significantly different from veterans of other wars or of interwar periods with regard to government support.

There were components of military experience not analyzed in prior research that were related to government support. When we converted our respondents' military occupations to their civilian counterpart occupations, we found that there was a positive relationship between social status of civilian counterpart occupation and support of the government. The mean score for personnel in military managerial jobs was 1.18. In office and clerical jobs it was .40. Skilled occupations averaged −.23, and operatives −.17. Personnel in service occupations averaged −.52 in government support. One interpretation of these data is that persons who moved to prestigious jobs within the military expressed gratitude for their status by supporting the government. An equally plausible interpretation is that this relationship reflects a positive correlation in the civilian labor force between status and government support, and jobs in the military merely reflected their incumbents' civilian status. The fact that soldiers who had been reduced in rank (N = 10) expressed significantly less support than those who had not likewise had at least two plausible explanations. They might be expressing resentment, or, alternatively, their alienation might have precipitated their demotion.

Soldiers who claimed to have found at least minimal application of their military skills in their civilian jobs had significantly higher government support scores ($\bar{x} = .40$) than those who claimed no such application ($\bar{x} = -.40$). Only three respondents indicated that they had jointed the service to learn a skill, and these three were right at the sample mean on government support, significantly lower than those who had enlisted for patriotic reasons ($\bar{x} = .56$).

Components of military experience had even weaker relationships to isolationism-interventionism than they did to support for government. Not surprisingly, conscripts were significantly less interventionist than persons who entered the service by more voluntary means (recruits, ROTC, service academy graduates). The former averaged –.26 and the latter –.05 on a scale ranging from –3 to +3. We also find that those who served in higher-status occupations were less interventionist than those in lower-status occupations. This is contrary to Schuman's findings (1972) on civilian public opinion regarding the Vietnam War, where those in higher-status occupations had more hawkish attitudes.

Impact of Service on Socioeconomic Status

As noted above, our veteran and nonveteran samples were found not to differ in social background characteristics. Tests for differences in race, religion, father's occupation, and age were all insignificant. However, in terms of their own achievements, the veterans were of significantly ($p < .05$) higher status than the nonveterans, using chi-square tests, although Kolmogorov-Smirnov tests did not quite reach significance at $p < .05$. In education, 33% of the nonveterans had not graduated from high school, as compared with 19% of the veterans. This is, in part, an artifact of the armed forces' minimizing the number of non-high school graduates who are accepted for service.

When we looked at family income, we found 14% of the nonveterans and only 2% of the veterans reporting incomes of less than $5,000. At the other end of the distribution, 51% of the veterans, and 46% of the nonveterans, reported incomes of $15,000 or more.

It was in comparing the occupational distributions of the two groups that we found the possible intervention of another variable: 23% of the veterans were managers, proprietors, or owners, as against 14% of the nonveterans; and, overall, 43% of the veterans and only 31% of the nonveterans were in white-collar occupations. However, 14% of the nonveterans indicated that they were retired, as against 6% of the veterans; and 11% of the nonveterans were students or unemployed, as against 4% of the veterans. Thus, nonveterans were less likely to be in the employed labor force than veterans and were more likely to be in nonemployed roles at the extremes of the age distribution. Some of the difference in employment, education, and income, however, is likely owing to utilization of veterans' benefits, for 28% of our veterans reported that they had taken advantage of GI educational benefits.

The overrepresentation of nonveterans in student and retired categories suggested an age difference. And although we had not found a significant difference in the mean ages of the two groups, the variance of age for the nonveterans was significantly greater than for the veterans ($F = 2.37$, $p < .01$), reflecting more people at the extremes. We therefore removed the students and the retired from the sample, reducing the difference in age variance between the two groups to insignificance. We then repeated our comparison of the two groups in terms of education, income, and occupation. Under these constraints, looking only at persons in the labor force (whether employed or not), there was no significant difference between the veteran and nonveteran groups in either occupation or family income. With regard to education, veterans were more likely to have graduated from high school than nonveterans. This difference was significant by chi-square test, but only approached significance at the .05 level using a Kolmogorov-Smirnov test. Significant or not, any difference is understandable in terms of the avoidance by the military of non-high school graduates, the opportunities for completing secondary school while in the military, and the availability to veterans of GI Bill benefits to further their education.

Discussion

Contrary to much polemical writing on the effect that the Vietnam War had on those who fought it, but consistent with recent research on Vietnam veterans, we find that those Detroiters who have ever served in the military are no different from those who have never served. Although we have too few cases to draw substantive conclusions, we also find that our veterans from the Vietnam era tend not to differ from veterans of other wars or of interwar periods in this regard.

When veterans and nonveterans were compared, the most striking finding was the absence of differences between the two groups. Similarly, when the effects of differing experiences within the military were analyzed, most dramatic was the extent to which there were not many differences—at least with regard to the two attitude dimensions with which we were concerned.

We did replicate the Bachman and Jennings finding on the relationship between branch of service and government support, perhaps manifesting the traditionalism of the Navy and the non-traditionalism of the Air Force, reflected either in self-selection or in socialization of our respondents. And we found, not surprisingly, that enlistees saw the U.S. military as playing a more active role in world affairs than did conscripts. This is perhaps a manifestation of self-selection for military service. We did not find, however, that military service or its components made the veterans in our Detroit sample any more or less interventionist, or supportive of or opposed to the way the federal government conducts its business.

An analysis of Vietnam veterans in the Project TALENT sample has recently produced data indicating that military service did not have a negative impact on the subjective aspects of the quality of their lives relative to their peers who did not serve in the military (Wilson et al., 1975). While the authors of that report suggest that Army service had a positive effect for the veterans, their data show no significant differences between veterans and nonveterans. Our data likewise show minimal differences. Despite potential effects of self-selection and/or adult socialization within the military institution, the lives of veterans seem not, on the whole, to have been touched hard by military experience in the areas we have studied.

This is not to deny that the Vietnam War, and previous military engagements, were times of great stress and reorientation for many who served—and for many, the costs of service persisted long after the wars were over. We do not disagree with Helmer and others who demonstrate that there are veterans who bore great burdens as a result of the Vietnam War and other wars. Indeed, many gave their lives. We seek here merely to place observations of the extreme effects of military service in the perspective of the average response. That the modal effect of service seems to be no effect at all does not belie the fact that the personal consequences of such service are measured on continua, and cases exist at the extremes as well as in the center of the range. Indeed, extreme cases are unlikely to be included in a household sample survey.

VI.

PATTERNS OF CIVIL-MILITARY RELATIONS

THE PUBLIC VIEW OF THE MILITARY

John D. Blair and
Jerald G. Bachman

Public attention was increasingly focused on the U.S. military during the course of the Vietnam War. Many negative aspects of the military and its personnel were highlighted. These included financial scandals among senior enlisted men, blatant mistreatment of recruits, the specter of widespread drug use, and the nightmare of My Lai type atrocities of unknown dimensions. These revelations and the failure to achieve military supremacy despite massive expenditures of resources (both men and material), the increased outrage over the war and the draft expressed on campuses and in mass rallies and marches, and the growing disaffection with the war evidenced in the polls led many to believe that the military had encountered a "crisis of legitimacy" in which it had lost its support among the public. In other words, it was felt that the disaffection with the war and what it revealed about the military had generalized into disaffection for the military.

It is time to reassess this assumption about what has happened to the public view and assessment of the military. Howard Schuman (1972), in an insightful analysis of the "Two Sources of Antiwar Sentiment in America," demonstrated empirically that the growing dissent concerning the Vietnam War might be only tangentially related to the sorts of arguments being advanced in the antiwar movement—especially as expressed on university campuses. He distinguished "moral" opposition to the war from "pragmatic" dissent based on the frustrating expenditure of vast resources coupled with only very meager results to be found in that particular war. Therefore, growing disaffection for Vietnam did not, for most people mean increasing moral rejection of the war, but probably reflected growing pragmatic opposition to a costly disappointment.

AUTHORS' NOTE: *The data for this paper were collected by the Survey Research Center and the Center for Research on the Utilization of Scientific Knowledge of the Institute for Social Research, University of Michigan. Portions of the research were supported by the Office of Naval Research and the Army Research Institute for the Behavioral and Social Sciences. The views expressed herein are those of the authors and do not represent institutional positions of the Department of the Navy or the Department of the Army.*

Hence, the generalization of antiwar sentiment to antimilitary sentiment found on campuses and reflected in the destruction of ROTC facilities and/or the removal of ROTC from full academic participation might well not be reflected in mass public attitudes toward the military, where antiwar sentiment was based primarily on pragmatic considerations rather than moral ones.

In fact, a number of studies at the Institute for Social Research (which asked respondents to rate various groups on a "feeling thermometer" ranging from 0° to 100°, with 50° as the neutral point) have found that the mean level of feeling toward the military has been warmer, or more positive, than for most other societal groups investigated; moreover, this level showed minimal decline from 1964 to 1972 (D. Segal, 1975a) and had returned to very near 1964 levels by 1974 (Inglehart and Barnes, 1975). A 1973 ISR study had respondents rate a number of institutions on how good a job they were doing for the country as a whole. Some of the findings have been reported in Rodgers and Johnston (1974). The U.S. military, along with the nation's colleges and universities, headed the list of 15 institutions. About 60% rated the military as doing a good or very good job, while only 10% rated it poor or very poor. In the same study, the people running the military were given one of the highest ratings for being "honest and moral." And when asked whether the military should have more influence in society, or less, over half the respondents said it should be "the same as now," and the rest were split almost exactly between those preferring more military influence and those preferring less. No other institution in the study got such evenly balanced ratings of influences.

The relative good ratings of the military in the 1973 ISR study were also found in a recent survey conducted by Louis Harris and Associates (1973) for the Senate Subcommittee on Intergovernmental Relations. In that study 40% said that they had "a great deal of confidence" in the military. The only institutions rated higher were television news (41%), higher educational institutions (44%), and medicine (57%). The several branches of the federal government, major companies, organized labor, and even organized religion all received lower confidence ratings than the military. However, the Harris study also indicated that in 1966 almost all the institutions had substantially higher ratings. In the case of the military, the proportion who expressed a great deal of confidence fell from 62% in 1966 to 35% in 1972, and then rose slightly to 40% in 1973.

In short, it appears that dissatisfaction with the Vietnam War did not always lead to a general antimilitary posture, although the influence of the war cannot be ignored. There is, however, considerable evidence that the secular trend of trust in government has paralleled that of support for the Vietnam War; i.e., both have decreased remarkably throughout the period from 1964 to the early 1970s (A. Miller, 1974). This is not to say that other factors–such as the civil rights movement, urban riots, recessionary periods, inflation, unemployment, and finally Watergate–have not been equally or perhaps even more involved in the decline of trust in government. It is, however, to say that pragmatic disaffection for the war might have been directed less toward a military whose hands were dirtied in an *immoral* war and more toward the government which was perceived to have been responsible for an *unwon* war. (See Modigliani, 1972, for a distinction between attitudes about "interventionism" and attitudes dealing with "administration distrust" found among the public during the Korean War.)

Another issue involved in the assumed generalization from revulsion with the horrors of war to antimilitary feelings is the effect of the war on its veterans. The Vietnam Veterans Against the War were eloquent spokesmen for the point of view that antiwar sentiment generalized into antimilitary sentiment, but again one must be careful in extrapolating to other Vietnam era veterans. Two separate longitudinal studies which included young veterans do not support such an extrapolation (Jennings and Markus, 1974; Bachman and Jennings, 1975). The point, therefore, is that it is time to reassess public views of the military at the end of the Vietnam era.

Data Sources

This paper addresses the basic question of what the public view of "the military" is. The findings are based on survey data collected from two samples: (1) a representative national cross-section of 1,855 civilians age 16 and older and (2) a sample of 2,522 Navy personnel stratified so as to be representative of major Navy entities (ships and shore stations). Sixteen-page, self-completed questionnaires, identical except for certain personal background measures, were administered to both sample during late 1972 and early 1973. A detailed description of sampling techniques as well as a description of the fit of the samples to their respective populations has been provided in an earlier report (Michaelsen, 1973). This paper extends and elaborates several analyses reported elsewhere (Bachman, 1973, 1974; Blair, 1975).

Interrelationships Among Military Values, Preferences, and Perceptions

This section of the paper is concerned with the ways in which values, preferences, and perceptions about military matters are interrelated. We begin by outlining the process of data reduction and describing the resulting dimensions. Then we explore some of the ways in which these dimensions are intercorrelated, looking separately at a number of different analysis groups in the Navy and among civilians. We review evidence to support our conclusion that the scales are interrelated in basically similar ways for the several groups. It should be added, however, that there are substantial differences in the way that groups are positioned along some of the scales; we review some of these differences later.

THE MEASURES OF VALUES, PREFERENCES, AND PERCEPTIONS

The questionnaire segment dealing with military values, preferences, and perceptions (Section C) includes 57 items, designed to measure a considerable number of different, but interrelated, concepts. An important phase of our analysis involved the consolidation of these items into a smaller number of indexes in order to produce multi-item variables, which are generally more stable and reliable than single items, and to reduce the complexity of the material to a more manageable level.

An early stage in our efforts toward data reduction involved a number of factor analyses including nearly all the items in Section C of the questionnaire. These analyses confirmed most of our prior expectations about sets of variables to be combined into indexes; in a few other cases, the analyses enabled us to locate items which did not meet our expectations.

These analyses were conducted separately for civilians, the Navy sample taken as a whole, and three subgroups within the Navy sample (officers, first-term enlisted men, later-term enlisted men). The patterns of factors which emerged from these several groups were quite similar; thus we felt confident that the indexes we were developing were applicable across all the groups examined.

The measures of values, preferences, and perceptions concerning military service which resulted from our data reduction efforts are shown in the left hand column of the tables and figures, although in Table 1, the separate perceived and preferred measures are excluded. As a matter of convenience, the measures are organized in the table according to the conceptual categories followed in our preceding papers.

Most of the measures are indexes based on two or more items. Three one-item measures are included because they are conceptually important but do not lend themselves to combination into indexes. Of the 57 items in Section C of our basic instrument, 42 are included in the 17 measures.[1]

A word is in order concerning the names given to the measures. An effort was made to capture the essence of an item or index in relatively few words, while at the same time conveying a good deal of the meaning. Some of the measures were better suited to this effort than others; in a few cases the names may seem a bit strained. In all cases, the name corresponds to a high score on the measure.

A GENERAL FACTOR OF PROMILITARY SENTIMENT

Our earlier explorations of the data, and some examination of the correlation matrices described above, led us to feel that there is a "general factor" of promilitary (or antimilitary) sentiment underlying most of the measures that we have been discussing. In an effort to test this notion we performed a set of factor analyses.

As a first step, product-moment correlations were computed among all of the measures in Table 1. The correlations were computed for each of the following analysis groups separately:

Navy first-term enlisted men
Navy later-term enlisted men
Navy officers
Civilian men
Civilian women

The complete correlation matrices are presented in Bachman (1974).

Table 1. LOADINGS ON A GENERAL FACTOR OF "PROMILITARY SENTIMENT"

Measures of Military Values, Preferences, and Perceptions	Factor Loadings* For: Navy Sample: 1st-Term Enlisted Men (N = 1,194)	Navy Sample: Later-Term Enlisted Men (N = 834)	Navy Sample: Officers (N = 310)	Civilian Sample: Men (N = 753)	Civilian Sample: Women (N = 1,053)
The Military Work Role					
Perceived military job opportunities	.6088	.5830	.7212	.4225	.4030
Perceived fair treatment in services	.5727	.5665	.6997	.5015	.4982
Perceived discrimination against women and blacks	–.2142	–.3715	–.4677	–.3996	–.3772
Military Leadership					
Perceived competence of military leaders	.6909	.6491	.7739	.6661	.6924
Military Influence Over National Policy					
Preference for higher military spending and influence	.6542	.4486	.7707	.7270	.6951
Role of military in society perceived as negative**	–.5786	–.3236	–.5351	–.5421	–.5778
Adequacy of military influence (perc. minus pref.)	–.5225	–.3295	–.6630	–.6274	–.5149
Foreign Policy and Military Power					
Support for military intervention	.4572	.4945	.5689	.4127	.2626
Preference for U.S. military supremacy	.5810	.4368	.6479	.6032	.5886
Vietnam dissent	–.7195	–.6928	–.7749	–.6919	–.6832
Issues Involved in an All-Volunteer Force					
Support for amnesty	–.5855	–.6045	–.7273	–.6464	–.6067
Opposition to unquestioning military obedience**	–.5508	–.5079	–.4293	–.5607	–.5710
Opposition to obedience in My Lai-type incident**	–.5232	–.3235	–.3289	–.5398	–.4823
Preference for "citizen-soldiers" (vs. "career men")	–.0804	–.3197	–.4203	–.1301	–.1203
Preference for wide range of views among servicemen	–.0854	–.2852	–.3019	–.3515	–.2025
Variance explained (by first factor)	28.4%	23.1%	36.1%	29.5%	26.5%

*Table entries are loadings on the first factor (unrotated) resulting from factor analyses using the principal components method.

**These measures are single items. All others are indexes based on two or more items.

In this series of factor analyses we were looking for the largest and most general single factor underlying the military value, preference, and perception measures. Accordingly, we used the principal components method and focused attention on the first factor (unrotated). The factor loadings for each of the five analysis groups are displayed in Table 1.

The results shown in Table 1 clearly confirm our view that there is a rather substantial general factor of "promilitary sentiment" which contributes to our measures of military views. It accounts for or "explains" between 23% and 30% of the variance in these measures for Navy enlisted men and civilians. It accounts for 36% of the variance for Navy officers, and the factor loadings for this group tend to be somewhat higher than is true for other groups. This appears to be a reflection of education, for we have shown similar factor loadings for civilian college graduates with 40% of the variance explained.

There is a considerable degree of similarity in the patterns of factor loadings for all five analysis groups. Without exception, the direction of loading is the same for all analysis groups–i.e., a measure is either positively loaded for all groups or negatively for all. Moreover, those measures which load most strongly are the same across all groups.

Let us consider what it meant (in early 1973) to be high in our general factor of promilitary sentiment. Not surprisingly, those highest in promilitary sentiment rated our military leaders as quite competent, gave the military services high markes for job opportunity and fair treatment, stated a preference for higher levels of military spending and influence, and saw the role of the military in society as predominantly positive. Their foreign policy views were rather "hawkish"–they were relatively supportive of U.S. military intervention in other countries, they preferred a position of military supremacy (rather than parity with the U.S.S.R.), and they were most likely to support past U.S. involvement in Vietnam. Finally, they placed a high value on obedience to military authority–they tended to agree that "servicemen should obey orders without question," and some maintained this position even when faced with a My Lai-type incident.

Two measures which show little association with the general factor of military sentiment are the dimensions most closely linked to the debate about the draft versus the all-volunteer force; these measures are Preference for Citizen-Soldiers (versus "Career Men") and Preference for Wide Range of Political Views among Servicemen. These two dimensions seem to stand somewhat apart from most of the other measures and are less integrated into an overall promilitary or antimilitary continuum. Our basic conclusion is that our respondents showed little "polarization" along these dimensions–perhaps indicating that most people had not given much thought to the issues they represent.

Among all the dimensions summarized above, the measure of Vietnam Dissent has a particularly strong loading on the general factor of military sentiment. One possible interpretation for this relationship is that those generally supportive of the military establishment were, as a result, least critical of our past involvement in Vietnam. In other words, Vietnam views were shaped by broader attitudes about the military. An alternative interpretation is that views about the Vietnam involvement were generalized to the larger military establishment, so that negative feelings about Vietnam led to negative views about military spending, influence, leadership, and the like. These two interpretations are not mutually exclusive–indeed, it is likely that both patterns of causation were at work.

Returning to the basic question of the nature of the public view of the military, we conclude that Vietnam sentiment is, in fact, closely related to the overall view of "the military." More generally, we have found that most of our items distribute respondents along a promilitary-antimilitary continuum; i.e., individuals are similarly ordered from one item to another. But does this consistent set of relationships among an individual's attitudes mean that the anti-Vietnam War sentiment reported in the polls has "generalized" to other aspects of the military? It is to an examination of the distributions of the different items that we turn next for an answer to this question.

The Public View of the Military

The first portion of this paper stressed the intercorrelations among our measures of military views–a pattern indicating that individuals who were most favorable toward the military along one dimension tended to be relatively favorable along other dimensions. But this does not mean that for a typical individual–or for the public in general—views about the military were consistently positive or consistently negative. To the contrary, there were substantial differences in overall ratings of the several aspects of the military that we explored; along some dimensions the dominant public view was quite favorable, while along others it was not.

In this section of the paper we will look first at mean scores on the various measures for all civilians in our sample—the public as a whole. Then we will turn to some variation in the responses to the measures resulting from differences in age and education. Finally, we will examine the effect of personal experience with the military on the values, perceptions, and preferences of civilian men.

THE PUBLIC AS A WHOLE

In Figure 1 we have summarized the basic findings for the total civilian sample. The mean score on each measure is reported in terms of its relationship to the substantive midpoint.[2] This relationship is determined both by the differences of the sample mean from the midpoint and by the standard deviation of the total civilian sample on that item. Hence if a mean score were +.25 scale points different from the midpoint and the standard deviation were .50, then on the figure, the bar would extend to the first line (1/2 S.D.) on the "promilitary" side. The end of the bar represents the location of the mean score with respect to the substantive midpoint or division between promilitary and antimilitary responses expressed as proportions of the standard deviation for the sample as a whole. This rather complex procedure was used because the measures differ considerably in terms of the lengths of the scales used as well as the variance from one scale to another, and we wanted to increase the comparability across scales.

In Figure 1 we find that although the items are highly interrelated and display a common underlying factor, their distributions do not demonstrate that in early 1973 the public was primarily either pro- or antimilitary. The public *as a whole* evaluated different aspects of the military quite differently, even though each measure may order *individuals* similarly.

Our analysis leads us to believe that there are three fundamentally different aspects of the military that are evaluated by the public and reflected by these findings. The first of these deals with the *military organization* itself. The scales labeled "the military work role" and

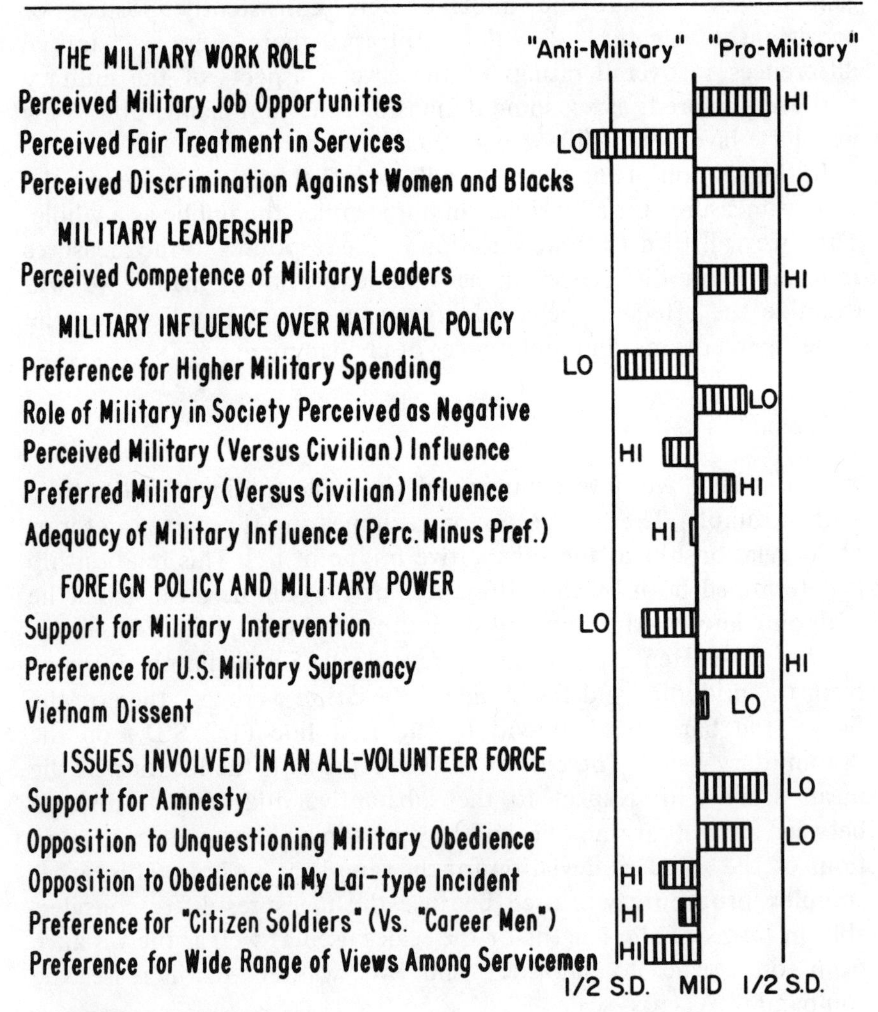

Figure 1: MEAN SCORES OF ALL CIVILIANS CONTRASTED WITH MIDPOINT OF EACH MEASURE

"military leadership" and the item dealing with "unquestioning military obedience" all deal with dimensions of the military as an organization: how competently it is managed, what it provides, how its members are treated, and what can be expected of the members. Generally this aspect of the military is positively evaluated. On the other hand, its members are not seen as having much recourse if unfairly treated—especially in comparison to that enjoyed by civilians.

The second fundamental aspect of the military deals with the use of *military force.* The measures found under the label "foreign policy and military power" as well as the one asking about "obedience in a My Lai-type incident" reflect attitudes about the appropriate conditions for going to war, the necessity of maintaining military superiority, and the outcome of the Vietnam War. The findings suggest that the mass public is reluctant to go to war except to protect the United States from actual attack, but that it will support efforts to deter such an attack through a high level of military superiority.

Our measure of Vietnam dissent did not show in early 1973 the overwhelming rejection of the war generally reported in poll findings. We feel that this is a reflection of our not just asking respondents the common question of whether the war was a mistake, but asking instead a series of questions that contain both "moral" and "pragmatic" considerations and that combine issues of intervention and national defense. Unfortunately these measures were developed long before Schuman's seminal work appeared (1972), and the moral and pragmatic aspects of sentiment toward the war were not clearly delineated in our questions. The end result of the combination of the various items is to provide a measure that indicates ambiguity in the public view of the Vietnam War (Bachman, 1973).

The third fundamental aspect of the military concerns *civil-military relations.* The measures we see as reflecting values, perceptions, and preferences in this area are found under the heading "military influence over national policy" and the three items not yet examined under "issues involved in all-volunteer force." Civil-military relations involve an interchange between the civilian and military sectors of society or, expressed somewhat differently, between society as a whole and the military organization.

How is the military's relationship to society evaluated? Our data reveal that the role of the military in society is seen as quite positive.

In addition, although the military is seen as having considerable influence vis-à-vis civilians in areas dealing with national security and internal organizational concerns, the public prefers it that way, and hence the influence of the military is seen as generally adequate —neither excessive nor inadequate.

These findings come from data collected about 6 months earlier than those that provided the basic findings reported in the *Washington Post* by Chapman (1974) and later by Rodgers and Johnston (1974). The summary of the data provided by the *Washington Post*'s headline, "Military Most Admired US Institution," certainly overstates the case, but our findings support the basic thrust of Rodgers and Johnston's findings, which are that the military organization and the role it plays in society are still positively evaluated by the "average person." Although our data cannot speak directly to the question, the consistently high levels that the military received on the thermometer scores (reported above) throughout the entire Vietnam era lead us to believe that the findings coming from these two studies are not a reflection of a change in the public image of the military after the war, but indicate a continuing high level of support for *the military organization.*

The other side of the coin in *civil-military relations* is what society does for the military. Here the picture is somewhat mixed in terms of public attitudes. There is a considerable reluctance to provide more money for the military.[3] In addition, there seems to be some reluctance about having only career soldiers or limited political views within the military. These last two items, as noted above, are important issues in an all-volunteer force but do not seem to be highly tied to other ideas about the military or to reflect very well-defined attitudes on the part of most people.

The last item dealing with civil-military relations is the index dealing with amnesty. Here we would like, primarily, to interpret its consequences for the view of the relationship of society to the military. The rejection of amnesty in the findings seems to us to indicate the basic legitimacy of military service even in an agonizing and divisive war. It must be remembered that our measure of attitudes toward the war itself revealed considerable belief in its importance for national security, although it was seen as harmful to the country generally.

In summary, three points need to be emphasized:

1. The distributions of the responses to the various measures we

have used reveal that, although similar ordering of individuals may exist across the measures, there remained considerable differences in the values, perceptions, and preferences of the public as a whole concerning different aspects of "the military."

2. At this aggregate level, civilians did not appear entirely or even primarily pro- or antimilitary. There was a generally favorable evaluation of the performance of the military organization and its relationship to society, but a dissatisfaction with the present high levels of financial support for the military and a reluctance to support the use of military force except in self-defense.

3. The differences in attitudes concerning the military organization and those concerning the use of the force made possible by that organization led to a clarification of apparently discrepant prior research findings. Some of these studies indicated that the public rejected Vietnam policy and acts of atrocities, but others showed that the public positively evaluated military leaders and the job done by the military for the country.

VARIATION BY AGE AND EDUCATION

Thus far we have looked at the distribution of respondents' values, perceptions, and preferences only at the aggregate level. During the Vietnam era, Americans became sensitized to the counterculture expressed among youth and to the quite different attitudes displayed by hard hats. These reactions to the Vietnam War and to "the military" suggest that aggregate findings do not reflect adequately some very important differences in the public view of the military. There are other possible bases of cleavage within the civilian population that may also be important in examining these issues. However, our other work with these data reveals that the primary bases of cleavage are age and education.

For simplicity in presentation and comparison we have dichotomized age into "younger" and "older," defined as 34 or younger and 35 or older. In addition, we have separated our sample into college graduates and noncollege graduates. Our other work cited above has shown that these two background variables have generally consistent, linear relationships to the attitudinal measures that we have been examining here. As a result, little is lost in our understanding of these relationships by using dichotomies.

The findings for the four groups that result from combining these

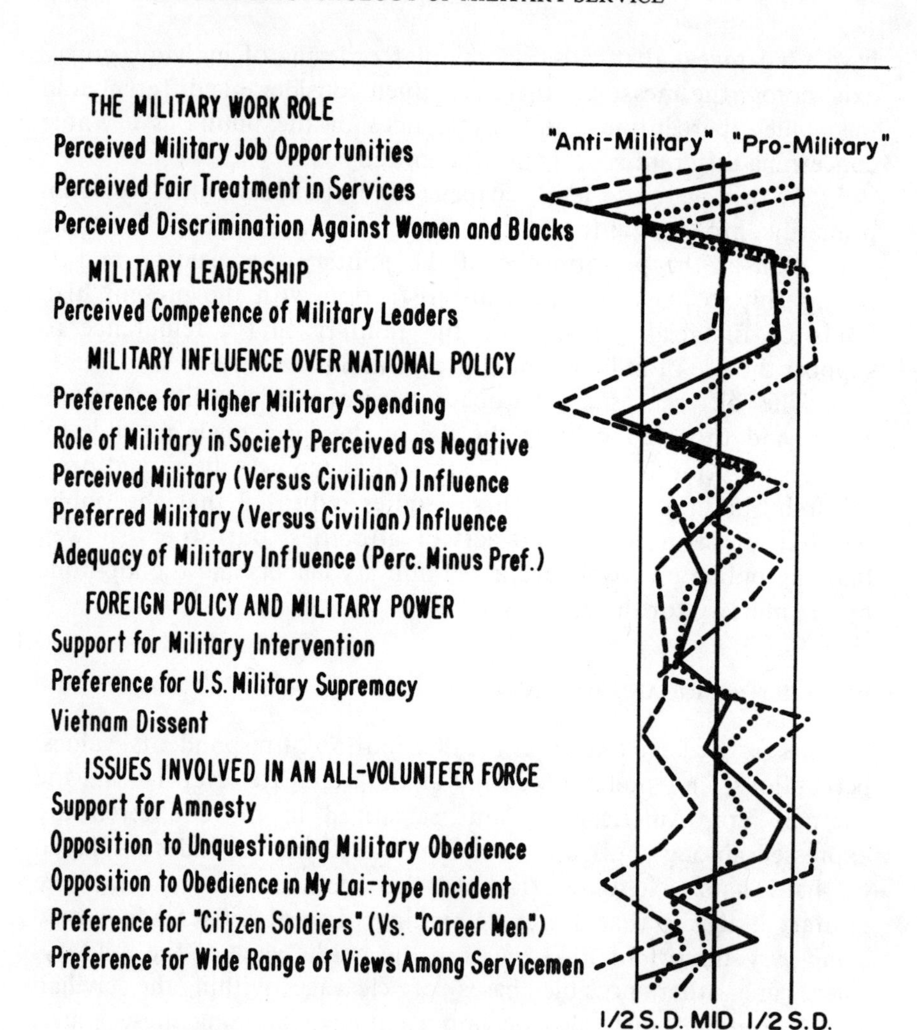

Figure 2: MEAN SCORES OF CIVILIAN GROUPS CONTRASTED WITH MIDPOINT OF EACH MEASURE

two dichotomies are shown in Figure 2. The same measurement and data manipulation procedures are used as in Figure 1. Because the variance from one group to the next *on each item* is very similar, we have again used the standard deviation of the civilian sample as a whole, and the results are reported in proportions of standard deviation units.[4]

It should be particularly noted that younger college graduates showed a fairly high level of "antimilitary" responses. This is consistent with Schuman's finding (1972) that moral, as opposed to pragmatic, opposition to the Vietnam War as discovered in his sample was found primarily among *young* college graduates in the mass public. However, even here the major distinction between the military organization and the use of military force applies. The young graduates were consistently more opposed to *all* forms of using military force than they were to the military organization, with the exception of the measure of fairness within the military services. They also evaluated civil-military relations in a very negative light. They wanted considerably reduced military spending, perceived a very influential military whose influence is excessive, and were concerned about all the issues involved in an all-volunteer force.

The older nongraduates showed the most consistently "pro-military" views. But their consistency was less than that of the younger college graduates. They consistently perceived the military organization as a positive entity (except, again, for fair treatment). When it came to the use of military force, however, they were also reluctant to support any action not related to national defense. In fact, their scores on the support for military intervention index equal those of the younger college graduates and probably reflect two very different types of isolationism. In terms of civil-military relations, the older nongraduates saw the military as making an important contribution to society and were less reluctant to provide money and to support greater than present military influence. Nor did they seem greatly concerned about the kinds of issues raised by the all-volunteer force.

The other two groups fall in intermediate categories in terms of pro- or antimilitary attitudinal consistency. Some aspects of the military were more likely to be called into question by older college graduates, and other aspects were more critically evaluated by the younger nongraduates.

In sum, there were, in early 1973, important variations within the

Table 2. IMPACT OF PAST MILITARY SERVICE ON ATTITUDES OF CIVILIAN MEN

	Correlations* With:	
	A. Past Military Service** (All civilian men—N = 719)	B. Positive Feelings About Having Served*** (Veterans only—N = 349)
The Military Work Role		
Perceived military job opportunities	.00	.18
Perceived fair treatment in services	.02	.33
Perceived discrimination against women and blacks	−.06	−.22
Military Leadership		
Perceived competence of military leaders	.02	.31
Military Influence Over National Policy		
Preference for higher military spending and influence	.07	.33
Role of military in society perceived as negative	−.05	−.34
Perceived military (versus civilian) influence	−.23	−.17
Preferred military (versus civilian) influence	.07	.25
Adequacy of military influence (perc. minus pref.)	−.23	−.30
Foreign Policy and Military Power		
Support for military influence	.10	.18
Preference for U.S. military supremacy	.07	.18
Vietnam dissent	−.07	−.32
Issues Involved in an All-Volunteer Force		
Support for amnesty	−.19	−.36
Opposition to unquestioning military obedience	−.04	−.35
Opposition to obedience in My Lai-type incident	−.09	−.29
Preference for "citizen-soldiers" (vs. "career-men")	−.16	−.08
Preference for wide range of views among servicemen	.06	−.10

*The correlations are product-moment; however, the "past military service" variable is a dichotomy, and thus the correlations in Column A are also termed point-biserial. (The point-biserial is a special case of the product-moment correlation; in the present instance it can be interpreted in essentially the same way as the more typical product-moment correlation involving continuous distributions along both dimensions—see Nunnally, 1967, pp. 120-133.) Correlations in Column A may be considered statistically significant if they exceed .11, those in Column B if they exceed .16 ($p < .001$, two-tailed, assuming a simple random sample).

**Past Military Service is based on responses to Question D14, but scoring has been reversed so that a positive correlation means that veterans have higher scores on the dimension than do nonveterans.

***Positive Feelings About Having Served is based on responses to Question D20, with scoring reversed so that a positive correlation means that the dimension is associated with positive feelings about have served.

public values, perceptions, and preferences concerning the military. The most important of these were linked to the age and education of the respondents. At the aggregate level, the younger college graduates exhibited the greatest consistency of attitudes. This may well reflect the generalization of concerns with the Vietnam War to all things military, although even in this group the military organization fared better than did the use of military force or the state of civil-military relations.

VETERANS' VIEWS OF THE MILITARY

There are a number of ways of forming attitudes about the military. But firsthand personal contacts may have a much greater impact and credibility. Thus it seemed useful to explore the degree to which the various perceptions and attitudes about the military and its mission are linked to such personal contacts.

In this section we will explore firsthand military contacts among the civilian men in our sample. First, we will see to what extent civilian men who had served in the armed forces differed from those who had not. Second, we will explore the extent to which positive or negative feelings about past military service were related to military attitudes and perceptions.

Is the average veteran more supportive of the military than the average nonveteran? Our data, summarized in column A of Table 2, indicate few differences between the average veteran and nonveteran along our dimensions, and those differences which do appear are not very large. When asked how they would feel about a son's enlistment, 29% of the veterans checked "strongly positive," compared with 23% of the nonveterans; those responding "mostly positive" were 45% and 46%, respectively. Differences this small are neither statistically trustworthy nor substantively important (they correspond to a point-biserial correlation of .08).

Reviewing the figures in column A of Table 2, we find no average differences between veterans and nonveterans in their perceptions of the military work role or the competence of military leaders, and no substantial differences in their views about foreign policy and military power. In ratings of ideal levels of military versus civilian influence, there was little difference between the two groups; both veterans and nonveterans preferred a roughly equal sharing of influence by military and civilian leaders. On the other hand, there was a difference in perceptions of actual levels of military influence; veterans perceived the military as somewhat less influential than the nonveterans. As a result, the two groups differed along our dimension of adequacy of military influence; veterans preferred a bit more military influence than they felt was actually the case, whereas nonveterans preferred slightly less.

It is interesting to note that the veterans' scores along the military versus civilian influence dimensions were quite similar to those of first-term enlisted men in the Navy. (See also in this volume Bachman and Blair's " 'Citizen Force' or 'Career Force'? Implications for Ideology in the All-Volunteer Force.") It may be that one of the most consistent results of past or present experience in military service is a lowered assessment of the amount of influence that military leaders actually have over a range of decisions affecting national security.

Veterans and nonveterans differed little in their evaluations of past U.S. actions in Vietnam; both groups were split nearly equally between those who tended toward support and those who tended to be critical. In their feelings about amnesty, on the other hand, the groups differed noticeably. While 36% of the nonveterans agreed or agreed mostly that the men who went to Canada to avoid fighting in Vietnam were doing what they felt was right and should not be punished, only 18% of the veterans agreed or agreed mostly. In answer to the comparison question stating that going to Canada was wrong and those who did so should be punished, 78% of the veterans agreed or agreed mostly, compared with 61% of the nonveterans. It is perhaps understandable that most civilians who once served in the armed forces themselves would have little tolerance for those who avoided service by going to Canada.

One other difference is worth noting between veterans and nonveterans. The veterans showed greater support for the idea of "career men" in the military rather than "citizen soldiers," whereas the opposite was true for nonveterans. The data are summarized in Table 3.

Table 4 displays two questions which deal with veterans' reactions to their military service and their perceptions of family reactions. Most veterans reported their own feelings as positive; nevertheless, there was room for variation. This variation in feeling about one's own military experience is strongly associated with other attitudes about military matters.

Table 3. PREFERENCE FOR "CITIZEN-SOLDIERS" VERSUS "CAREER MEN"

Scores on the Index	Veterans (N = 324)	Nonveterans (N = 386)
1.0-2.0 Prefer "career men"	46%	31%
2.5 Mixed feelings	24	26
3.0-4.0 Prefer "citizen-soldiers"	30	43
Total	100%	100%

Perceptions about family reactions, shown in the lower part of Table 4, were more balanced between positive and negative views. The two items shown in Table 4 were only modestly correlated (r = .23); this finding, coupled with the different levels of positive feeling shown in the two items, indicates that some veterans held positive feelings about their military experience in spite of a perception that their entry into the service was not especially favored by family members.

Veterans' own feelings about having served are strongly correlated with feelings about the possibility of a son's enlistment (r = .53). These feelings about having served are also related to a number of other dimensions, as shown in column B of Table 2. The pattern of correlations is similar to the factor loadings shown in Table 1; those items which are most positively or negatively associated with a general factor of promilitary sentiment are also most strongly linked to veterans' feelings about their own military experience.

In column B of Table 2, veterans' positive feelings about their military service are correlated with perceptions that the military organization offers fair treatment and competent leadership.

Table 4. VETERANS' FEELINGS ABOUT HAVING SERVED

Question D20. Would you say your feelings about having been in the military are:

(1) Strongly positive	38%
(2) Mostly positive	43
(3) Mostly negative	14
(4) Strongly negative	5
	100%

Question D21. Which of the following best describes the feelings of your family when they first learned you were going to enter the service?

(1) They were very much in favor of it	20%
(2) Somewhat in favor	26
(3) Neutral or indifferent	25
(4) Somewhat dissatisfied	22
(5) Very much dissatisfied	8
	101%

Veterans with positive feelings about their past service were also likely to prefer fairly high levels of military influence, view the role of the military in our society as being predominantly positive, support past U.S. actions in Vietnam, show above average opposition to amnesty, and state that servicemen should obey orders without question.

We explored several other dimensions of veterans' experiences to see if they correlated with our measures of military attitudes. No substantial or consistent differences were found to be associated with branch of service (Army, Navy, Air Force, Marines). There also appeared to be no clear differences in attitudes between those veterans who had been drafted and those who had enlisted–although most of the veterans who had enlisted thought they would have been drafted otherwise.

One dimension which did show some consistent differences is length of past military service. Most of our civilian respondents who served in the military remained in the service no more than 4 years. But those who had served longer tended to show attitudes more favorable to the military. In particular, the greater the length of service among these veterans, the higher their ratings of fair treatment and competent leadership in the military. These findings are probalby a reflection of career orientation among the longer-serving veterans.

In summary, the effect of personal contact with the military was not found to be as strong as is sometimes supposed. This might well indicate the differences between studies of large and representative samples as opposed to those focusing on case studies of individuals or organized groups such as the VFW or the VVAW (see Helmer, 1974). There was, of course, enormous variation in attitudes within the group of veterans–as was true for the nonveterans as well. Some of this variation among veterans can be explained in terms of differing evaluations of their own military experience. However, a recent study of personal contact with governmental agencies (see Katz et al., 1974) found that only *negatively* evaluated contact has any effect different from no contact at all. Similarly, we found consistent relationships between the evaluation of military experience and military attitudes but little relationship between veteran status and attitudes. Most veterans were positive in their evaluation, as shown in Table 4; so it appears that those with negative contact were negative in their views, but those with positive contact did not have greatly different scores from those with no contact. Had the evaluations of past military service been more evenly distributed, the lack of differences in mean scores between veterans and nonveterans could have resulted from an averaging process, but that is not the case here. Those who had negative experiences also had negative attitudes; those who had positive experiences looked like those with none at all–i.e., like nonveterans.

Summary and Conclusions

In this look at the public view of the military, we have stressed several different themes.

First, we noted a fairly strong pattern of intercorrelations among the different dimensions along which people rated the military and its mission. In general, we found that those who were most favorable toward the military along one dimension tended to be among the more favorable–or less critical–along other dimensions.

The second finding–that the military and its mission received "mixed reviews" at the hands of a national cross-section of civilians–may help shed light on some apparently discrepant prior research findings. Several studies found that the public rejected Vietnam policy to an increasing degree during the late sixties and early seventies, but other research indicated that the public fairly consistently gave high ratings to military leaders and to the job that the military was doing for the country. We have argued that these two kinds of findings are not incompatible, particularly if the government–rather than the military leadership–was blamed for getting the U.S. involved in Vietnam. At the same time, it must be acknowledged that those most strongly critical of Vietnam policy also tended to be the least supportive of the military along other dimensions, thus suggesting that for some individuals a frustration with Vietnam policy may have led to a heightened dissatisfaction with the military as a whole.

Evaluations of the military and its mission vary somewhat across different groups. Younger college graduates as a group not only called the use of military force into question but were also critical of virtually all aspects of the military organization and existing civil-military relations. On the other hand, older nongraduates as a group showed predominantly promilitary attitudes, but were still critical of the use of military force for intervention, high levels of military spending, and the fairness of treatment within the military. The other two groups examined—older college graduates and younger nongraduates—exhibited intermediate levels of pro- or antimilitary attitudes. The older graduates were more likely to call some aspects of the military into question, and the younger nongraduates to evaluate others critically. When we compared veterans and nonveterans, we found few differences in attitudes about the military. The veterans were, however, more likely to rate the military as less influential, were more strongly opposed to amnesty, and were more supportive of a "career military" than were nonveterans. Within the group of veterans there were variations in ratings of the military: those who were dissatisfied with their own service experience tended to be less favorable to the military as a whole.

Notes

1. The 17 measures contain one instance of redundancy. The indexes of perceived Military Influence and Preferred Military Influence are ingredients for a single discrepancy measure (Perceived minus Preferred) which indicates the extent to which a respondent thinks the level of actual military influence exceeds, or falls short of, what he would consider ideal. In our factor analyses the separate Perceived and Preferred measures are excluded, thus leaving a set of 15 measures in which each item appears no more than once (Table 1).

2. The substantive midpoint is the point on the scale which divides responses into those that are favorable or "promilitary" and those that are unfavorable or "antimilitary." The classifications used are consistent with the results of the factor analyses reported.

3. Blair (1975) has separated the item (C25) dealing with military spending from the index combining preference for higher military spending and influence. This single item is the measure we have used in Figures 1-2. The two items used in the index are very highly correlated.

4. Generally, a difference between the two nongraduate groups of more than .20 standard deviation, between a nongraduate group and a graduate group of more than .25 standard deviation, and between the two graduate groups of more than .30 standard deviation is both substantively and statistically significant. Also, the location of the mean of either nongraduate group more than .10 standard deviation (and more than .20 standard deviation for either college graduate group) from the midpoint means that the distribution is not evenly divided but leans significantly in the pro- or antimilitary direction.

"CITIZEN FORCE" OR "CAREER FORCE"? IMPLICATIONS FOR IDEOLOGY IN THE ALL-VOLUNTEER FORCE

Jerald G. Bachman and
John D. Blair

Introduction

In the years of debate that preceded the return to an all-volunteer staffing of the armed forces, a number of problems or objections were raised. Some questions were concerned with costs, others with whether a sufficient number of volunteers could be obtained; perhaps the most profound set of issues centered on the societal and political effects of moving to an all-volunteer force. We will discuss below our concern about a "separate military ethos" resulting from a military force made up of career men rather than "citizen-soldiers." Such issues and problems have been discussed at length (see especially Tax, 1967; U.S. President's Commission on an All-Volunteer Armed Force, 1970) and we have also dealt with them in some of our other research (Bachman and Johnston, 1972; J. Johnston and Bachman, 1972; Bachman, 1973, 1974; Bowers and Bachman, 1974; Blair, 1975). One of our purposes in the present study has been to tap levels of public awareness and concern over some of these issues. The main purpose, however, has been to show some of the implications for the all-volunteer force if certain assumptions made by the President's Commission on an All-Volunteer Armed Force were faulty.

Recently Janowitz and Moskos (1974) demonstrated that the assumptions made by the President's commission concerning the racial composition of the all-volunteer force were greatly in error. The commission had expected the racial as well as other aspects of the demographic composition of the all-volunteer force to remain

AUTHORS' NOTE: *The data for this paper were collected by the Survey Research Center and the Center for Research on the Utilization of Scientific Knowledge of the Institute for Social Research, University of Michigan. Portions of the research were supported by the Office of Naval Research and the Army Research Institute for the Behavioral and Social Sciences. The views expressed herein are those of the authors and do not represent institutional positions of the Department of the Navy or the Department of the Army.*

quite similar to that of the "mixed force" made up of a mix of conscripts, draft-motivated volunteers, true volunteers, and a career force of reenlistees. Janowitz and Moskos presented data on the racial composition of the armed forces that show that a clear racial imbalance (compared to the population as a whole) was developing —contrary to the commission's expectations.

Our own concern is not with racial imbalance, but with the possibility of a "promilitary" ideological imbalance. This imbalance could result from increases in the proportion of career-oriented military personnel under all-volunteer condtions. Moskos (1970) has argued that there has been a change over time in the sources of strain and differentiation within the military from one between officers and enlisted men to one between career-oriented military men (both officers and enlisted men) and those who are not career-oriented.

The President's commission (1970) assumed that the makeup of the all-volunteer force in terms of career orientation would not differ greatly from that found in the mixed force. For example, it was believed that the turnover rate of first-term enlisted men would be about three-quarters of that found under nonvoluntary conditions; i.e., about 65% of the first-term enlisted men would not reenlist after their first tour of duty in an all-volunteer force. This would result in only a slightly larger proportion of the total force (48% compared to 40%) consisting of reenlistees that the commission referred to as the "career force."

Janowitz and Moskos (1974:112) report that "in fiscal year 1973, the reenlistment rate among first-term Army enlisted men was 52.0% for blacks compared with 35.1% for whites." The point we want to stress here is that even in the fiscal year *preceding* the official start of the all-volunteer force, Army enlisted men who had entered primarily as conscripts or draft-motivated volunteers several years earlier were already reenlisting at the maximum rate that the commission expected would exist in an all-volunteer force. Indeed, among black Army first-term personnel, the reenlistment rates considerably exceeded the 35% expected by the President's commission. Since reenlistment rates clearly affect the overall proportion of reenlistees versus first-termers, we conclude that the President's commission may have greatly underestimated the relative size of the career force which might develop under all-volunteer conditions.

In this paper we will argue that, as the proportion of career-oriented men in the all-volunteer force increases, that force will be less likely to match the values, perceptions, and preferences concerning the military held by civilians. To put it another way, our findings suggest that an enlarged proportion of career men will increase the danger of what has been called a "separate military ethos." (The sources of the data analyzed, the kinds of measures used, and the civilian views of the military are discussed in Blair and Bachman's paper "The Public View of the Military," presented earlier in this volume.)

PUBLIC VIEWS OF THE ALL-VOLUNTEER FORCE

The results from questionnaires and also from an interview segment administered only to civilians indicated strong majority support for the concept of the all-volunteer force and relatively little concern about some of the issues which have been raised as potential problems. The nationwide civilian sample supported the all-volunteer approach rather than the draft by nearly a two-to-one margin. There was also very strong support for the higher military pay levels considered to be necessary under a volunteer system.

When asked about issues related to the types of people who would staff the military services, there was a slight tendency for people to favor citizen-soldiers over "career men," but the views seemed rather mixed. Civilian responses to open-ended interview items about the all-volunteer force left a dominant impression that most people have not thought much about the question of what kinds of servicemen will, or should, staff an all-volunteer armed force.

The makeup of an all-volunteer force has been of considerable concern to us, however. The findings that we will present here are admittedly speculative, but represent an attempt to come to grips with an important question: to what extent are the attitudes found among military men representative of civilian values, perceptions, and preferences concerning the military—and is the answer for career military men different from what it is for noncareer men?

The findings for the total civilian sample and for young men aged 19 to 24 indicate a much more negative image of the military among young men than was found in the public as a whole (Figure 1).[1] Indeed, along every dimension the young men showed average scores below those for the total public, and many of the differences were quite large. Not surprisingly, the view of these men was most similar to that found among younger ($\leqslant$ 34) college graduates in our other

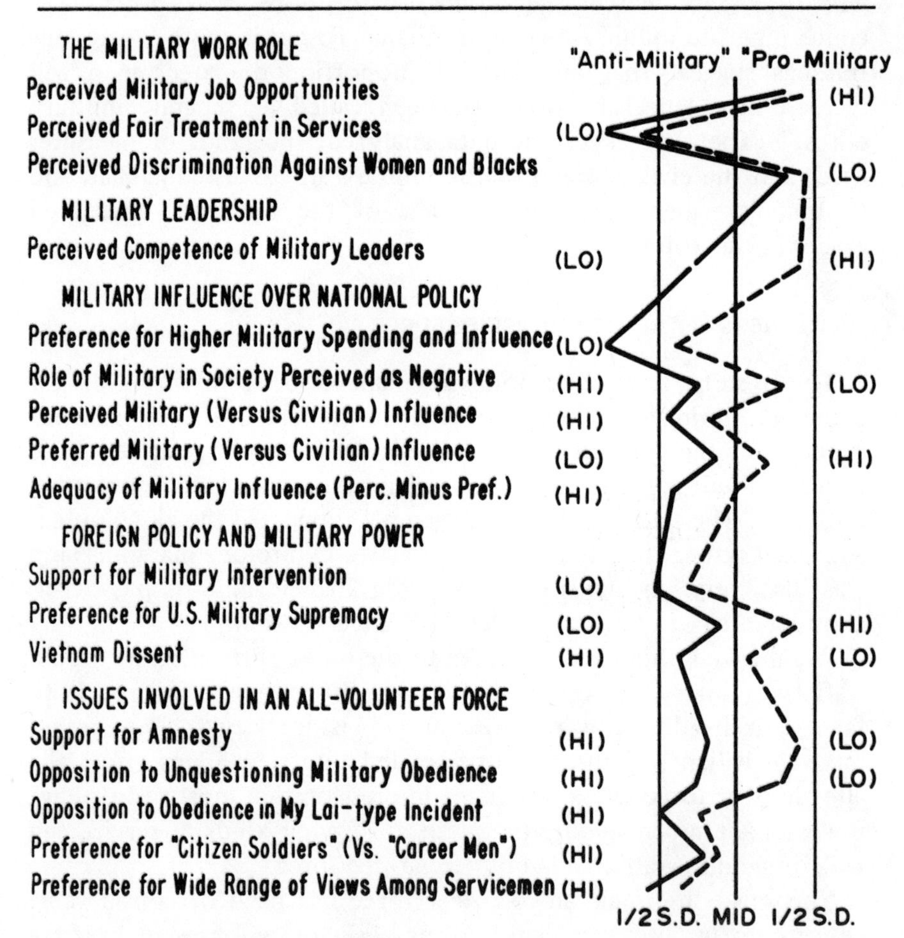

Figure 1: MEAN SCORES OF CIVILIAN MEN 19-24 AND ALL CIVILIANS

analysis. The majority of these men were not college graduates, but their younger age (19-24) seemed to capture the negative images of the military found in the counterculture.

The differences between these young men and the public as a whole included a rejection of the use of *military force* in Vietnam. (Young men tended to reject intervention in general, but this was also true of the population as a whole.) Among the young men there was even considerable ambivalence about whether the U.S. should have military superiority over other countries, including the U.S.S.R. Although there appears to have been some differentiation in the attitudes about the use of force for intervention and the use of force for self-defense, there was considerably less support for the use of military force of any kind than was found in the public as a whole.

In addition, the role of the military in *civil-military relations* was seen differently. More specifically, the role of the military in society was seen by the majority of young men in our sample as negative; and military influence vis-à-vis civilians was judged as excessive —primarily because it is *perceived* to be extensive. Considering the civilian side of civil-military relations, a very substantial majority also wanted to see military spending curtailed; and a majority called the legitimacy of military service during the Vietnam War into question and supported amnesty for those who refused to serve.

Even the *military organization,* which generally fared well in public attitudes, was seen in a less favorable light by the younger men, particularly in terms of fair treatment. This view was generally found in the civilian public, but to a slightly less extent. In addition, the competence of the organization's leaders and the value of unquestioning military obedience were disputed by the young men in our sample. On the other hand, the only aspects of the military positively evaluated by the young men in Figure 1 also deal with the military organization. It was perceived by the majority of young civilian men to provide good job opportunities and to have relatively low levels of discrimination against women and blacks.

Since this sample represents the most likely recruits for an all-volunteer force, the negative views they hold of the military point to considerable difficulties in recruiting from this group generally. This does not mean that there will be no recruits, for there have been—even before the present economic recession provided added impetus to enlistments. But it does raise an important issue: are those who serve voluntarily a balanced and representative cross-section of their age group? It is to this question that we turn next.

MILITARY VIEWS LINKED TO CAREER MOTIVATION

Our focus of attention will be young Navy men—specifically, first-term enlisted men. In examining the views of first-term enlisted men, we will separate them into the two groups of interest to

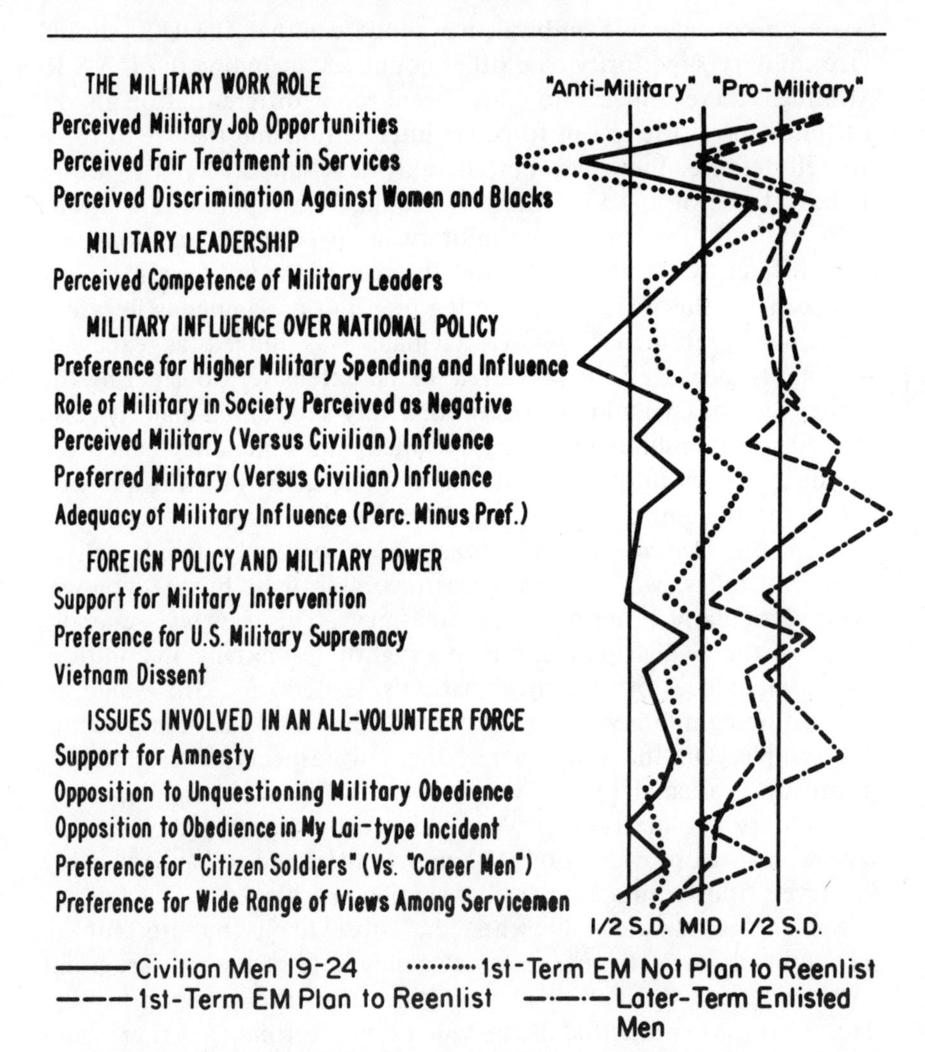

Figure 2: MEAN SCORES OF CIVILIAN MEN 19-24 AND THREE NAVY GROUPS

us—those who intend to reenlist and those who do not. In addition to comparing these two groups to civilian young men aged 19-24, we will also introduce another group for comparison purposes—later-term enlisted men. This last group corresponds to what the President's commission called a career force. One of the reasons for looking at later-term enlisted men is that they constitute the group most likely to be in positions of authority and direct supervision over incoming junior enlisted men. We expect that, to most first-term enlisted men, it is the more senior enlisted personnel—especially noncommissioned officers—who represent the military establishment and its values.

Figure 2 presents mean scores for first-term enlisted men who planned to reenlist and mean scores for those who did not, as well as mean scores for young civilian men and later-term enlisted men.[2] By far the most striking finding, seen across a wide range of our measures, is that the career-oriented first-termers were remarkably similar in ideology to their later-term colleagues, with virtually all scores averaging on the promilitary side of the ledger. On the other hand, those young men who planned to leave the military were rather similar to their civilian peers in most matters of ideology about the military—both groups tended to be critical of the military, sometimes quite strongly.

While the pattern described above dominates Figure 2, there are some distinctions worth noting. When we compare civilian young men with the first-termers who planned to leave the military, it appears that the first-termers were more critical of the *military organization*—perhaps reflecting their own dissatisfaction and eagerness to leave. The largest and most consistent differences involved the area of *civil-military relations;* the young Navy men who were viewing the military from inside—even though they were planning to leave—saw the role of the military in society as a bit more positive and were less critical of military spending and influence. Indeed, these Navy men preferred a slight increase in levels of military influence, whereas the opposite was true for the young civilian men, on the average. In a number of other respects, including views about Vietnam and amnesty, the noncareer-oriented first-termers were indistinguishable from their civilian counterparts.

Turning now to those first-termers who planned to reenlist, we emphasize that, in practically every area measured, these career-oriented young men were substantially different from their civilian peers and also from their first-term counterparts who planned to leave the military. The question naturally arises: why did the career-oriented young men in the Navy hold views so clearly different from their civilian age-mates and also from their service-mates who planned an early return to civilian life?

ATTITUDE CHANGE OR SELF-SELECTION

In our earlier analyses of these data, we considered two alternative explanations of the findings outlined above:

(1) During the first tour of duty, those individuals most likely to reenlist may undergo *attitude changes* in a more promilitary direction. This may occur through a process of socialization as a result of exposure to the more experienced Navy men who tend to hold such views, or of exposure to positive experiences in the Navy, or both.

(2) By the time they reach their late teens, some individuals may be more favorable than others in their view of the military services and mission. These differences, which exist prior to enlistment, may be among the factors influencing the *self-selection process* involved in the decision to reenlist.

While the only really adequate test of these two competing explanations would involve a longitudinal design, we felt we could gain some insights by looking separately at first-termers who had served about 1 year, those who had served 2 years, and those who had served 3 or 4 years. If self-selection accounts for the differences between the attitudes of the career Navy men and others, there should be consistent differences in attitudes between those who did and those who did not plan on reenlistment—i.e., the differences for those in their first year should be just as large on the average as the differences found for those in their second, third, or fourth years of service. On the other hand, if the attitude change explanation is correct, we *might* expect to see smaller differences among those in their first year—assuming that the process of attitude change requires more than a few months to be completed.

Our basic finding was that the differences between first-termers who planned to reenlist and those who did not were evident quite early. Those who had served about one year showed differences just as large on the average as those who had served several years longer. This finding is fully consistent with the self-selection explanation —the view that reenlistment is heavily influenced by rather deep-rooted perceptions and ideology related to the military life style and mission. The alternative explanation, based on attitude changes during the first tour of duty, is not ruled out entirely. Indeed, both explanations could be true to some degree. But whatever the pattern of causation, our analyses in this area demonstrate that it does *not* require years and years of service experience for first-term enlisted men to develop the strongly promilitary attitudes found among later-termers. For those who planned to reenlist, the same attitudes were clearly evident as early as the first year of service.

Career Force, Citizen Force, or Mixed Force?

One of the arguments raised in the debate about the all-volunteer force was the danger of a separate military ethos brought about by a military force made up largely of career men. The findings summarized above suggest some basis for concern in this area. To the extent that new recruits into an all-volunteer force consist increasingly of the sort of career-oriented personnel we have been studying here, it seems inevitable that the military will indeed grow more separate from civilians—at least when it comes to views about the military and its mission.

In this section of the paper we will present a more speculative analysis. We will try to isolate the implications, for ideology, in the military, of different possible locations of the all-volunteer force on the continuum from citizen force to career force. We will treat our total Navy sample as a real-world example of a mixed force, while the career-oriented component of that sample will be treated as an example of what a career force could look like.[3]

The findings are presented in Figure 3. This figure differs from the other two in that we have not compared the two Navy groups to the substantive midpoints of the measures, but have contrasted them with the mean score of the civilian population as a whole on each measure. (In other words, the mean scores of the Navy groups have been charted according to the extent to which they differ from the total civilian sample.) The differences are expressed as proportions of the standard deviation for all civilians.[4]

COMPARING THE "MIXED FORCE" AND THE PUBLIC

Figure 3 reveals some very interesting things about the aggregate similarities and differences in the values, perceptions, and preferences of the total Navy—our "mixed force"—and civilians as a whole. In evaluating the military organization, we found that the total Navy was virtually identical to the civilian sample in perceptions of job

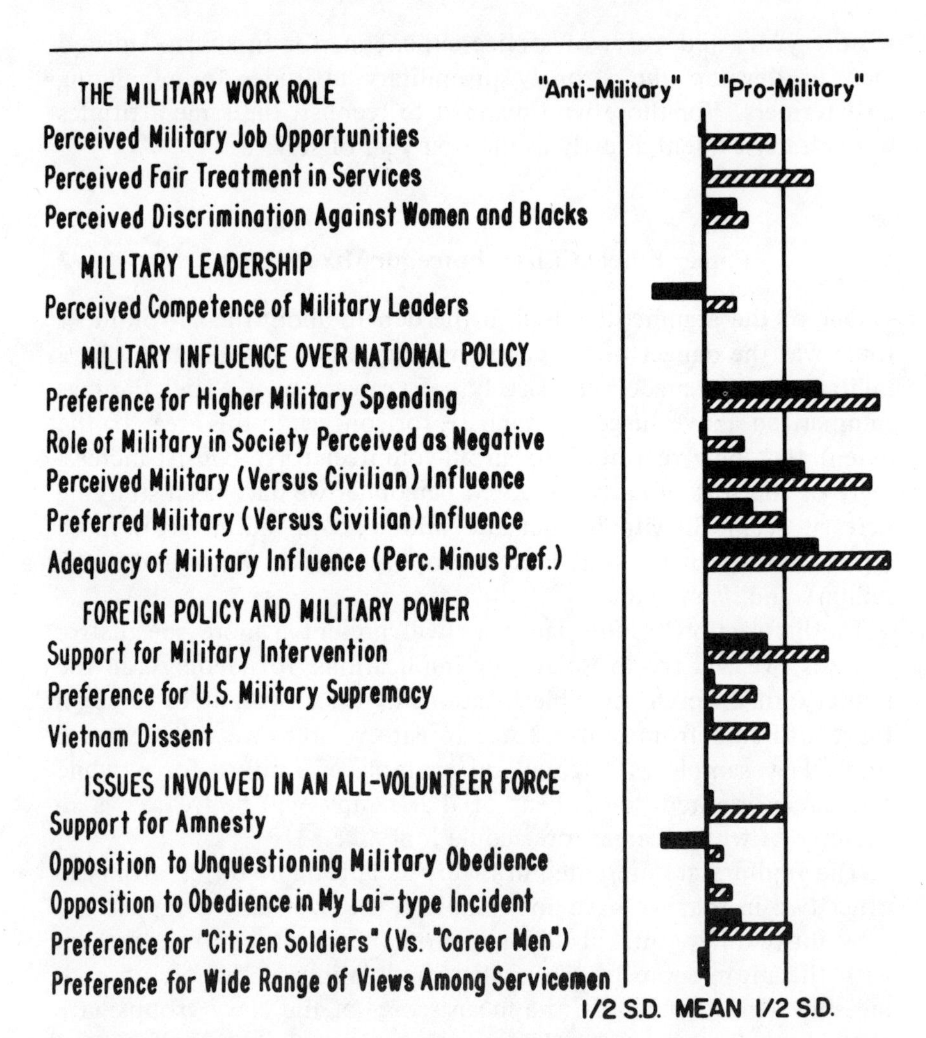

Figure 3: TOTAL NAVY AND CAREER NAVY CONTRASTED WITH MEAN SCORES OF ALL CIVILIANS

opportunities and fair treatment in the services and held a slightly more positive view of the level of racial and sexual discrimination within the military organization, but showed relatively lower satisfaction with the competence of military leaders and with the concept of unquestioning military obedience.

In terms of the use of military force, the only difference between the total Navy and civilian samples was a higher level of support for military interventionism among Navy men. Attitudes about military supremacy, the Vietnam War, and a My Lai-type situation showed no differences.

When we turn to issues of civil-military relations, we see considerable differences between the total Navy and civilian samples. Although the perception of the role of the military was the same, there were great differences in views about the level of money that society should provide for the military. On the other side of the issue, the Navy men were much more likely than civilians to rate the military as having little influence vis-à-vis civilian leaders. There was also more support for high military influence in the Navy sample than among civilians, although this difference was not as great as that dealing with perceptions of actual influence levels. These differences resulted in the view among military men that the influence of the military leaders is very inadequate. There was also more support, among members of the "mixed force," for having the military consist of career men rather than citizen-soldiers.

These comparisons show that the total Navy in early 1973 was rather similar to civilians as a whole in views about the military organization and the use of military force. Presumably, the considerable similarity between the total Navy and civilians reflects the "civilianizing" effect of large numbers of noncareer enlisted men and officers. The differences that did exist were concentrated primarily in the areas of civil-military relations.

COMPARING THE "CAREER FORCE" AND THE PUBLIC

The other comparison available in Figure 3 is that between the public as a whole and our representation of a "career force" as a pure type. Here we are concerned with examining what a force consisting only of career-oriented members would look like in terms of values, perceptions, and preferences concerning the military.

In virtually all the aspects of the military that we have examined, the career-oriented portion of the Navy sample was clearly more "promilitary" than the other groups examined. The members of the career military viewed all aspects of the military very differently than did the population as a whole. This was especially true for civil-military relations, as shown in Figure 3.

Thus we conclude that, to the extent that an all-volunteer force consists primarily of career-oriented men (and women)—and thus approximates in reality our career force—the attitudes found among its members as a whole will be very discrepant from those found among civilians as a whole. If this is the future of the all-volunteer force, it will be considerably less representative in this respect than was the mixed force of the past.

Policy Implications and Recommendations

This paper has documented a number of important ideological differences between career Navy men and their noncareer or civilian counterparts. We think these differences may have important implications for the all-volunteer force of the future.

Under present conditions, an all-volunteer force is likely to recruit and retain personnel from only part of the ideological range found in the civilian population. The very individuals who are needed to broaden the ideological balance are probably the least likely to enlist—or reenlist. Present conditions in the services are changing, and such changes may help to obtain a representative cross-section of volunteers. But if the nation's leaders value the concept of the citizen-soldier or citizen-sailor, they will do well to broaden the incentives in ways that are especially attractive to those presently underrepresented among volunteers. And, in spite of the additional costs involved, it would be wise to seek out some kinds of enlistees who are likely to serve for one term only and then return to civilian life.

WHAT KINDS OF RECRUITS AND HOW TO RECRUIT THEM?

Career Navy men—and those most likely to become career men—tended to be more zealous about the military than their civilian age-mates. This is one of the strongest and most consistent findings in this paper. There is much to indicate that these differences were due, at least in part, to processes of selection–the more promilitary were likely to reenlist in the Navy. These findings on reenlistment, which held true for a Navy cross-section in late 1972, are more and more likely to apply to first enlistments, now that we have an all-volunteer system.

How should military recruiting efforts respond to this finding that its enlistees and especially its career men are likely to come from only a limited ideological range? One approach is to embrace this state of affairs enthusiastically, recognizing that the more promilitary individuals are likely to be less troublesome and more in agreement with traditional military values and practices than some of their less gung ho contemporaries. Indeed, the idea of concentrating recruitment efforts on those most favorably disposed toward the military is one of the specific recommendations in a recent report to the Army that introduced the concept of the "quality man"–an individual who, among other things, says that he places high importance on patriotism, is proud of being an American, would be among the first to defend the country if it were attacked, and is generally more favorable toward military service (Opinion Research Corporation, 1974).

The approach of aiming recruitment efforts toward the more gung ho is understandably tempting to recruiters and perhaps to many others in the military. And it may appear to be successful in the short run. But in our view such a recruiting approach would be unwise in the long run. It would tend to reinforce and heighten the tendencies we have already observed for career military men to be less than fully representative of the cross-section of civilian viewpoints. By strengthening support for some unnecessary and perhaps counterproductive military traditions and practices–or at least reducing resistance to them–this approach could gradually widen the gap between the military and the civilian worlds. We suspect that this gap would eventually reduce the supply of recruits below an acceptable level. Such a gap would also increase the risk of developing a separate military ethos.

An alternative approach, the one we recommend, is to develop recruitment efforts designed to obtain a broader and more fully representative cross-section of individuals among first-termers and also among career personnel in the military. The primary advantage of such an approach is that it tends to avoid the problems and pitfalls mentioned above. An additional advantage is that extending recruiting efforts beyond the most gung ho may help to attract some of the brightest and most ambitious young people to a period of military service. (We have found that promilitary attitudes are somewhat *negatively* related to education, and findings in other studies of young people suggest that those who go to college are more likely to express critical views of the military *in its present form.* Thus an effort to increase recruiting among those presently more cautious about military service is likely to involve some of the most able of our young adults.)

How could the military services go about implementing this approach of seeking a broader and more representative cross-section in its recruits? Two types of strategy may be distinguished, and we recommend both. First, the extrinsic incentives to enlistment—those rewards or inducements which are not directly linked to actual performance in the work role—should be geared toward a broader cross-section of individuals, especially those who have relatively high educational abilities and interests. The second strategy is to modify intrinsic characteristics of military work roles so as to make them more broadly attractive. Elsewhere we have offered a number of specific recommendations for improving Navy work roles and effectiveness (Bowers and Bachman, 1974).

Probably the most obvious extrinsic incentive that comes to mind when considering any work role—military or civilian—is pay. The higher the level of pay, the more attractive the work role is assumed to be. Certainly in discussions about the feasibility of converting to an all-volunteer force, primary attention was directed toward increasing military salaries; efforts were made to estimate how much money would be required to induce enough men to enlist under volunteer conditions (President's Commission on an All-Volunteer Armed Force, 1970). The recent pay increases were surely a *necessary* condition to the establishment of an all-volunteer military force, but in our view the higher salaries do not constitute *sufficient* conditions—and in some respects the emphasis on pay increases may have led us to overlook other important incentives to military service.

EDUCATIONAL INCENTIVES TO ENLISTMENT

One set of incentives which is worth greater attention and emphasis is the educational benefits available to servicemen during and after their tours of duty. Although the young men (and women) bound for college represent a group especially high in ability and ambition, military recruiting policy has to a large degree treated them as unlikely prospects (Binkin and Johnston, 1973). And in its recent report to the Army, the Opinion Research Corporation (1974, p. viii) advised that, "While college students do not express strong opposition any longer to the military as an institution, enlistment still does not appeal to them. Noncollege men remain the Army's major market." But in that same report it is noted that educators rate "interference with education" as a primary deterrent to military service and feel that this drawback could be offset by greater emphasis on the GI Bill as a source of support for a college education.

In sum, under present conditions the typical high school student planning for college tends to view military service as an unwise interruption of his educational development. Given no change in present conditions–or, worse yet, given any reduction in educational benefits for veterans–it is probably accurate to conclude that noncollege men will remain the primary source of military personnel. But we think it would be unwise to leave present conditions as they are. On the contrary, *we recommend that the educational benefits available to those in service and to veterans be retained and enhanced and that these benefits be publicized more widely.* In particular, we would suggest the establishment of specific "pay your way through college plans" that stress the opportunity to qualify for veterans' benefits, amass substantial savings, and accumulate some college credits during a tour of military service following high school.

But why should the military deliberately seek out individuals who are likely to serve only one term and then go on to college as civilians? Some of the advantages in terms of high ability levels and broader perspectives have been noted above, and these help to balance out the costs of higher turnover among those who enter the military in order to work their way through college. But it should be added that a considerable degree of turnover is necessary and desirable in an organization that has only limited positions of leadership at the top. The "college in exchange for service" formula is a means of attracting able individuals who can learn quickly, serve effectively, and then leave to make room for fresh recruits. Moreover, high rates of turnover among this group of individuals would not be a sign of organizational failure, and those who left would not be spending their final years of service frustrated and disillusioned because the military had not lived up to their expectations. We agree with Friedman (1967) that some proportion of "in-and-outers" is desirable in the military services, and we view the use of educational incentives as a particularly effective means for ensuring this sort of turnover.

The "college in exchange for service" approach need not require that military service precede college. On the contrary, there would be substantial advantages for some to complete college and then enter the service. This would help meet military needs for skilled and educated personnel. Moreover, it seems likely that the broadening and liberalizing effects of higher education, plus the maturity of additional years, would make the college graduates less malleable, more confident and self-reliant, and better able to handle responsibilities than those recruited at an earlier stage of education and maturity.[5]

It is gratifying that the idea of increased use of educational incentives, which was supported by our earlier work (J. Johnston and Bachman, 1972) and reinforced by the findings presented here, has also been put forward by Janowitz and Moskos (1974) as one of the approaches for reducing racial (and social class) imbalance in the military. It is fortunate indeed that educational incentives can potentially deal with these problems while at the seme time helping to ensure—voluntarily—a mix of "in-and-outers" along with career personnel that is closer to a citizen force, not an ideologically isolated career force.

Notes

1. See "The Public View of the Military" in this volume for the format used in presenting data in Figure 1. For tabular data underlying this figure, see Bachman (1974), Tables 5 and 6.

2. The data in Figure 2 follow the format used in Figure 1 (see note 1).

3. Of course, the extent to which an all-volunteer force will empirically approximate a career force is unknown at this time. However, we are presently replicating this study with a sample from the all-volunteer Army. This should give us some idea of what proportion of current Army personnel fit the definition of "career-oriented" which we have used, and how that compares with our 1973 Navy sample, where 42% of the enlisted men and 72% of officers, or 46% of the total sample, were "career-oriented." (We have required only a stated intention to reenlist as the minimum definition of career orientation.)

4. If the end of the bar is more than .10 standard deviation from the middle line for the total Navy or more than .13 standard deviation for the career Navy, it indicates a

statistically significant difference between the mean of the Navy group and that of the total civilian sample.

5. We have discussed elsewhere a possible approach for offering GI Bill benefits in advance, coupled with loans, as a means of paying for college in advance of military service (J. Johnston and Bachman, 1972). Janowitz and Moskos (1974) have also suggested mechanisms for accomplishing this.

CHANGING VALUES AND ATTITUDES TOWARD MILITARY SERVICE AMONG THE AMERICAN PUBLIC

Ronald Inglehart

Events of recent years suggest that support for military institutions may be weakening among Western publics. In the United States and in a number of other Western nations, there have been indications of an increasing public reluctance to provide financial and moral support for the military. Such feelings seem particularly widespread among youth, who have sometimes shown not only reluctance but also refusal to serve in national defense forces.

Is this a transient phenomenon based exclusively on a response to specific issues at a given time, or could it reflect relatively deep-rooted trends in Western society? This is the question to be dealt with in this paper. The attempt to answer it raises complex analytic problems, and our findings mirror this complexity. On one hand, evidence that we will cite indicates that the great majority of the American public held surprisingly positive attitudes toward the military and military service in 1974; moreover, there is evidence that attitudes have become *more* positive in the immediate past. On the other hand, underlying this positive outlook, there may be a long-term change in basic values that tends to accord military security a declining priority among the goals most valued by the American and other Western publics.

The theoretical bases underlying the latter change have been outlined in an earlier exploration of the value priorities of six West European publics (Inglehart, 1971). We recapitulate our argument very briefly here. It was hypothesized (1) that people have a variety of needs and tend to give a high priority to whatever needs are in short supply, but (2) that they tend to retain a given set of value priorities throughout adult life, once it has been established in preadult years.

In the decades following World War II, Western publics experienced unprecedentedly high levels of prosperity and the absence of wars that threatened the physical safety of those being socialized in Western countries. Consequently, these publics have begun to give increasing emphasis to various needs *other* than those for economic and physical security. But basic values do not change overnight. Only among the youngest age cohorts—those who have been brought up since World War II—would we expect to find "postmaterialist" values widely prevalent.

The results of the earlier investigation tended to confirm these hypotheses. We found that among the publics of six West European countries surveyed in 1970, the younger age groups were far likelier to emphasize postmaterialist values; older groups—quite overwhelmingly—gave top priority to economic and physical security. Commenting on these findings, Samuel Huntington (1974) questioned whether the publics of postindustrial society would give sufficient emphasis to survival values—or whether they would even defend themselves in face of a genuine foreign threat.

The question may be pertinent. But before we can begin to deal with it, we need to know whether the sizable differences observed in the value choices of different age groups reflect life-cycle effects, generational change, or some combination of these and possibly other effects; only data collected over a period of many years can provide a conclusive answer. But it is interesting to note that the age-group differences were largest in countries, such as Germany, that have experienced relatively great changes in prevailing levels of economic and physical security; they were far smaller in countries, such as Great Britain, that have experienced relatively little change. If the age-group differences in a given country correspond to the amount of change that that country has experienced during the formative years of her respective generations, it strengthens the grounds for believing that these differences may be due to historical change rather than to life-cycle effects.

In 1972 and 1973 additional surveys were carried out in 11 countries, including the United States. Our expectation was that the American public would fall near the British end of the spectrum, in the strength of the relationship between values and age. We anticipated this because in recent decades the American public has (in comparison with the Germans, French, or Italians) experienced a relatively modest increase in prevailing levels of economic and physical security. Like Britain (only more so), the United States has had the advantages of geographic isolation, and it escaped invasion and devastation during the World Wars. But more recently, she has experienced relatively great foreign and domestic conflict. Until

1973, she was at war in Vietnam. The war, together with racial problems and a high crime rate, has contributed to domestic turbulence. During the formative years of her older cohorts, America was a haven of *relative* tranquility in comparison with most of Europe; today the positions seem to be reversed. In regard to physical security, there has been less difference between the formative experiences of America's older and younger cohorts —which should be reflected in a relatively small amount of value change across age groups.

The United States resembles Britain in another way: she was already *relatively* wealthy at the turn of the century, ranking far ahead of the other countries we are dealing with. Like Britain's (only less so), her postwar economic progress has been slower than that of other Western nations. In sum, one might expect the older American cohorts to be relatively postmaterialist—but the population as a whole should show less change across age groups than any of the European nationalities except, perhaps, for the British.

The data confirm these expectations. Table 1 shows the distribution of two "pure" value types in each of the 11 countries surveyed in 1972 and 1973, using a battery of four value-priorities items.[1] To simplify a complex table, only the two polar types are shown: the column headed "Mat." gives the percentage of materialists and the column headed "PM" gives the percentage of postmaterialists within each group (if one wishes to know the percentage falling into the mixed types, one can simply add up the figures for the two polar types and subtract from 100). All 11 countries show the same basic pattern that was found in 1970 and in 1971: the younger cohorts have a much higher proportion of postmaterialists and a lower proportion of materialists than the older cohorts. But the *rate* of change varies from country to country, in predictable fashion. The American sample shows less value change than any other country except Britain. The *oldest* American cohort has a higher proportion of postmaterialists than any European nation —reflecting the greatly privileged position this country once had—but the *youngest* American cohort has not moved toward postmaterialism as rapidly as some of their European peers.

In apparent value change, the Irish, Dutch, Danish, and Swiss form an intermediate groups—as we anticipated. And the Germans are at the high end of the scale, with the British at the opposite end, precisely as they were in 1970-1971. There is a remarkable stability in the relative position of given countries. The amount of value change seems to reflect a given nation's recent history.

Table 1. VALUE TYPE BY AGE COHORT IN 11 COUNTRIES, 1972-1973
(Original 4-item index, tabulated by age cohorts used in 1970 survey)

	Ages						Total Spread Between Youngest and Oldest Groups
	19-29	29-38	39-48	49-58	59-68	69+	
Germany							
Materialist	24%	39%	46%	50%	52%	62%	56 points
Postmaterialist	19	8	5	5	7	1	
France							
Materialist	22	28	39	39	50	55	51
Postmaterialist	20	17	9	8	3	2	
Italy							
Materialist	26	41	42	48	49	57	42
Postmaterialist	16	8	7	6	4	5	
Belgium							
Materialist	8	20	22	25	39	39	39
Postmaterialist	23	17	10	10	3	5	
Ireland							
Materialist	24	31	41	37	45	51	36
Postmaterialist	13	9	6	6	2	4	
Netherlands							
Materialist	27	22	28	40	41	51	35
Postmaterialist	14	17	9	10	12	5	
Denmark							
Materialist	33	34	47	44	48	58	34
Postmaterialist	11	9	4	5	4	2	
Switzerland*							
Materialist	27	26	30	35	34	50	32
Postmaterialist	15	17	15	9	6	6	
Luxembourg**							
Materialist	26		40		44		29
Postmaterialist	19		7		8		
United States							
Materialist	24	27	34	32	37	40	26
Postmaterialist	17	13	13	10	6	7	
Great Britain							
Materialist	27	33	29	30	36	37	17
Postmaterialist	11	7	6	7	5	4	

*Swiss data are from 1972; U.S. data are combined results of surveys in May 1972, November-December 1972, and March-April 1973.

**Because of the small size of the Luxembourg sample, it is broken down into only three age groups (ages 15-38, 39-58, and 59+).

Development of a More Broadly Based Values Indicator

Our four-item index seems to provide a measure of something pervasive and enduring in one's outlook. But we must not overlook this index's shortcomings. Probably its most serious weakness is the simple fact that it is based on only four items. Consequently, it may be excessively sensitive to short-term forces. For example, one of the items in the index concerns rising prices. Western countries have experienced extraordinary inflation in the past few years. It seems more than likely that the proportion of respondents giving high priority to "fighting rising prices" would go up—not as the result of fundamental value change, but simply because this is a very serious current problem. This type of instability would probably be much greater if we simply asked the respondents to rate the importance of rising prices *by itself;* but in our index, one's choice of this item is constrained by the fact that it must be ranked against *other* desired goals. Almost everyone was probably aware that rising prices were a more serious problem in 1973 than in 1970; nevertheless, by no means all of those ranked "freedom of speech" above "rising prices" in 1970 would be willing to change this ranking in 1973. The four items provide a better measure than one; but a more broadly based index would spread the risk over a still larger number of items, making it less likely that an individual's score would be unduly distorted by any particular recent event. Furthermore, a broader-based index might help reduce the amount of error in measurement —something which is always a major problem in survey research. In reply to survey questions, a substantial number of respondents seem to give superficial answers, more or less "off the top of their heads." With a single item, it is difficult to distinguish those whose answers reflect a genuine attitude from those whose response is essentially meaningless. But a set of consistent responses to a large series of related questions probably *does* reflect a genuine underlying preference.

In our 1973 surveys, we attempted to develop a broader indicator of an individual's value priorities. Analysis of the results should give us a more reliable measure of whether value change is taking place. It may also provide a more detailed picture of the respective world views of the materialist and postmaterialist types.

The 1973 survyes included the four items from the original value priorities index (see Card B below), but also included eight additional goals. The following questions were asked:

"There is a lot of talk these days about what the aims of this country should be for the next ten years. [HAND RESPONDENT CARD A.] On this card are listed some of the goals which different people would give top priority. Would you please say which *one* of these you, yourself, consider most important?"

CARD A

A. Maintaining a high rate of economic growth.

B. Making sure that this country has strong defense forces.

C. Seeing that the people have more say in how things get decided at work and in their communities.

D. Trying to make our cities and countryside more beautiful.

"And which would be the next most important? [HAND RESPONDENT CARD B.] If you had to choose, which one of the things on this card would you say is most desirable?"

CARD B

E. Maintaining order in the nation.

F. Giving the people more say in important government decisions.

G. Fighting rising prices.

H. Protecting freedom of speech.

"And which would be your second choice? Here is another list. [HAND RESPONDENT CARD C.] In your opinion, which one of these is most important?"

CARD C

I. Maintaining a stable economy.

J. Progressing toward a less impersonal, more humane society.

K. Fighting against crime.

L. Progressing toward a society where ideas are more important than money.

"What comes next?"

"Now would you look again at all of the goals listed on these three cards together and tell me which one you consider the *most* desirable of all? Just read off the one you choose."

"And which is the next most desirable?"

"And which one of all the aims on these cards is *least* important from your point of view?"

This series of questions enabled us to obtain relative rankings for 12 important goals. The introductory sentences placed the questions in a long-term framework, and the choices dealt with broad societal goals rather than the immediate needs of the respondent: we wanted to tap long-term preoccupations, not one's response to the immediate situation. The 12 options themselves were designed to explore the implications of a body of theory that seems highly relevant to the study of value change—Abraham Maslow's concept (1954) of a need hierarchy underlying human motivation. Six items were intended to tap emphasis on the physiological or materialist needs: "fighting raising prices," "maintaining economic growth," and "maintaining stable economy" being aimed at the sustenance needs, and "maintaining order," "fighting crime," and "maintaining strong defense forces" being aimed at the safety needs. The remaining six items, "giving people more say in government," "giving people more say on job," "progressing toward a less impersonal society," "protecting free speech," "making ideas count," and "making beautiful cities," were intended to tap various postmaterialist needs.[2] We view the latter needs as potentially universal: every human being has a need for esteem, an inherent intellectual curiosity, and a need for aesthetic satisfaction; one will act on these needs unless circumstances force one to stifle them. Our expectation, therefore, was that emphasis on the six materialist items would tend to form one cluster, with the postmaterialist items in another distinct cluster.

In order to test this hypothesis, we performed conventional factor analyses of the rankings of these goals in each of the ten countries.[3] The results show a cross-national uniformity which is almost breathtaking. In each case, five items—the *same* five items in every country—cluster near the positive end of the continuum. Six items—again, the same six in every country—cluster near the negative pole. The remaining item falls near the midpoint. This set of questions was replicated in another survey of the American public carried out in 1974.[4] The results of factor analyses of the two American surveys are shown in Table 2. The basic factor structure is virtually identical in 1973 and 1974—despite the fact that the questions were asked in a different format in the latter year.[5]

The items which cluster toward the negative pole are the six materialist items. And five of the six postmaterialist items fall into the opposite cluster. A single item—the one concerning "making more beautiful cities" (or "protecting nature from pollution" in the 1973 American survey)—does not fit into either group of items.

Table 2. THE MATERIALIST-POSTMATERIALIST FACTOR: REPLICATION OF 1973 SURVEY AMONG THE AMERICAN PUBLIC, 1974 (Loadings of value priorities items on unrotated first factor)

Goal	1973 (20%)	1974 (20%)
Ideas count	.508	.560
Less impersonal society	.627	.528
Freedom of speech	.409	.451
More say in government	.423	.438
More say on job	.451	.416
More beautiful cities	.278*	.119
Stable economy	−.435	−.392
Maintain order	−.491	−.409
Economic growth	−.397	−.432
Fight rising prices	−.334	−.449
Strong defense forces	−.464	−.467
Fight crime	−.484	−.502

*In 1973, this item as "Protecting nature from being spoiled and polluted."

Clearly, the people we interviewed did not respond to this question in the way that we had anticipated; but the other 11 items met expectations to an almost uncanny degree. The consistency of responses to these items cannot be attributed to such common sources of spurious correlation as response set: the questions were asked in a "cafeteria-style" format, which gives no cues to the "right" answer.

The empirical configuration of responses gave additional support to the hypothesis that these items tapped a set of hierarchically ordered needs. Given respondents tended to be preoccupied with a consistent set of needs located at either the materialist or post-materialist range of the continuum. Of the 12 items, 11 fell into two separate clusters, reflecting materialist and postmaterialist priorities respectively. The item designed to tap aesthetic needs fitted into neither cluster; with the same consistency by which the 11 other items *did* fit into their expected place, this one failed to show a single loading above the .300 level–not only in the United States, but in each of the 10 other countries surveyed. This interesting anomaly is discussed by the author (1976). Let us simply note here that 11 of the 12 items *did* seem to tap the dimension that they were designed to tap, permitting the construction of a more broadly based indicator of materialist and postmaterialist values. Let us note, furthermore, that a relatively high priority for "making sure that this country has strong defense forces" formed an integral part of the materialist cluster. We hypothesized that emphasis on defense against foreign enemies (like emphasis on maintaining order *within* the nation) tapped a more general physical security need–which should be linked with emphasis on economic security. The data confirmed this expectation.

The data also tended to bear out our suggestion that recent involvement in the Vietnam War may have heightened the perceived importance of military forces for the American public. This is a relative statement, however; it may be more accurate to say that a comparatively recent participation in military conflict simply retarded a decline of emphasis on this domain. For "maintaining strong defense forces" seems to be given relatively low priority today throughout the Western world. This goal was selected as first or second most important (our of 12) by very small minorities in Europe, ranging from a low of 2% in Belgium to a high of 7% in Italy. The United States ranked well above any European country –with 16% of the American public placing strong defense forces in first or second place. One could hardly describe the American public as intensely military-minded: "maintaining strong defense forces" got neither very high nor very low emphasis (for the sample as a whole, this goal ranked sixth out of 12 items). Yet, there is an impressive difference between the Americans and the various European publics, most of which have experienced continuous peace since 1945. We have little in the way of time-series data, but it seems almost certain that these publics must have placed much higher emphasis on national defense during World War II and its aftermath. If so, the current levels reflect decline–and the most obvious explanation for the distinctiveness of the American public would seem to be the fact that Americans have engaged in war much more recently than the Europeans.

Thus far, we have dealt with items designed to tap one specific dimension of an individual's value priorities–the materialist-postmaterialist dimension. It would be desirable to place these values in the context of one's *overall* value priorities. Milton Rokeach (1968, 1973) has developed a battery of items that seems highly promising for this purpose,[6] and we included Rokeach's Terminal Values Survey in our 1974 questionnaire. If it does, as designed, provide an exhaustive inventory of one's value priorities, some of its items should tap the materialist-postmaterialist dimension, enabling us to integrate our own work with his.

Table 3. VALUE DIMENSIONS AMONG THE AMERICAN PUBLIC: MATERIALIST-POSTMATERIALIST ITEMS PLUS ROKEACH TERMINAL VALUES
(All loadings above .250 in unrotated factor analysis based on cross-section of the American public, 1974)*

I. Materialist versus Postmaterialist Priorities (10%)		II. Peace versus Pleasure (8%)		III. External versus Internal Harmony (7%)	
Ideas count	.558	WORLD OF PEACE	.528	WORLD OF PEACE	.524
Less impersonal society	.481	NATIONAL SECURITY	.461	WORLD OF BEAUTY	.389
Freedom of speech	.456	EQUALITY	.439	EQUALITY	.373
WISDOM	.330	SALVATION	.347	COMFORTABLE LIFE	.308
WORLD OF BEAUTY	.316	Strong defense forces	.293	More beautiful cities	.271
More say on job	.284	FREEDOM	.279	FREEDOM	.255
HARMONY	.279	Maintain order	.252		
EQUALITY	.263				
More say in govt.	.255				
Maintain order	–.279	Fight rising prices	–.274	MATURE LOVE	–.285
Stable economy	–.283	MATURE LOVE	–.363	More say in govt.	–.293
FAMILY SECURITY	–.362	COMFORTABLE LIFE	–.440	SALVATION	–.295
Fight rising prices	–.401	HAPPINESS	–.472	Stable economy	–.345
Economic growth	–.415	EXCITING LIFE	–.504	WISDOM	–.414
Fight crime	–.436	PLEASURE	–.513	HARMONY	–.452
NATIONAL SECURITY	–.476			SELF-RESPECT	–.472
COMFORTABLE LIFE	–.480				
Strong defense forces	–.504				

IV. Beauty and Order versus Personal Achievement (6%)		V. Achievement versus Salvation (5%)	
More say in govt.	.503	SENSE OF ACCOMPLISHMENT	.545
More say on job	.451	Stable economy	.338
SENSE OF ACCOMPLISHMENT	.335	EXCITING LIFE	.256
Economic growth	.270		
FREEDOM	.263		
Fight rising prices	–.272	TRUE FRIENDSHIP	–.353
SALVATION	–.281	More say in govt.	–.398
Maintain order	–.353	SALVATION	–.435
Fight crime	–.375		
WORLD OF BEAUTY	–.387		
More beautiful cities	–.479		

*Rokeach items appear in all capitals. The figures in parentheses are the percentage of the total variance explained by the given factor.

Table 3 shows the results of a factor analysis of the 12 materialist-postmaterialist items, plus the 18 items in Rokeach's Terminal Values Survey. As we would expect, the configuration of an all-inclusive set of values is far more complex than the one generated by a set of items specifically designed to tap a single dimension. The strength of the avergae loading on the first dimension drops substantially, and the percentage of total variance explained by this dimension is cut in half. Yet the first factor is still clearly recognizable as the materialist-postmaterialist dimension: all 11 of the items that were linked with this dimension in Table 2 show relatively high loadings on the first factor in Table 3. Seven of the Rokeach items also correlate with this dimension, and in every case, the item relates to the given cluster in the way that we would anticipate on the basis of face content. In the materialist cluster we find emphasis on "a comfortable life" and "family security"–two concerns having an obvious linkage with economic security. Also located in this cluster is "national security"–further confirmation that this dimension has a direct relevance to one's attitudes toward national defense. In the postmaterialist cluster we find emphasis on "equality" and "inner harmony" plus a pair of items relating to the intellectual and aesthetic needs: "wisdom" and "a world of beauty." We had hypothesized that emphasis on beauty should load on this dimension. Ironically enough, neither of the items that we designed to tap the aesthetic needs showed the expected empirical relationship–but the item developed by Rokeach *does.*

Table 3 shows several other interesting dimensions, which we will not discuss here. By design, everything was thrown into this analysis. A more coherent first dimension emerges from analysis of Rokeach's 18 items with relatively strong loadings on the corresponding dimension in Table 3, as Table 4 indicates.

We have progressively enlarged the array of items empirically linked with the materialist-postmaterialist dimension, and in doing so have revealed its relevance to concern for military security. But we have moved a long way from the four-item index that was the basis of Table 1; do the amplified versions of the materialist-postmaterialist dimension show similar relationships with age?

The answer is yes–only more so. In our 1974 data, the 4-item index has a product-moment correlation with age of –.113; the factor scores generated by the 12-item analysis have a correlation of –.258; and those generated by the first factor in the 30-item set have

Table 4. THE MATERIALIST-POSTMATERIALIST FACTOR IN THE UNITED STATES, 1974 (Loading of 18 items on unrotated first factor)*

Goal	Loadings (16%)
Ideas count	.585
Less impersonal society	.506
Freedom of speech	.457
WORLD OF BEAUTY	.318
WISDOM	.317
EQUALITY	.302
More say on job	.300
More say in government	.274
INNER HARMONY	.268
Maintain order	–.280
Stable economy	–.302
FAMILY SECURITY	–.367
Economic growth	–.419
Fight rising prices	–.420
NATIONAL SECURITY	–.433
Fight crime	–.446
COMFORTABLE LIFE	–.461
Strong defense forces	–.496

*Rokeach items appear in all capitals.

a correlation of –.298. In broadening the base of our values indicator we have not diluted the relationship with age; we have strengthened it substantially. The possibility of intergenerational change in attitudes toward military service seems at least as strong as ever.

Does this imply that in the coming years we will witness a steady erosion of support for military service? In a tidy little monocausal world, it might–but the real world is hardly ever monocausal. Individual-level values are only one of the influences of public attitudes; system-level events, such as war or peace, the *type* of war or peace one experiences, prosperity or recession, the quality of a nation's leadership, and many other factors should certainly have some impact on mass perceptions and attitudes. Furthermore, it is still premature to say whether the relationship we have observed between age and the priority accorded to national security reflects a life-cycle effect, generational change, or some combination of the two.

One may even question whether these items really *do* measure values: perhaps they are simply a reflection of current fads, which the young pick up more readily than the old. If our "values" items actually *do* tap values, we would expect them to be relatively resistant to short-term changes. We do have some evidence on this score, from both Western Europe and the United States.

Early 1970 was a period of high prosperity and full employment throughout Western Europe. There had been an almost unbroken economic expansion for two decades. The years 1971-1973 were a period of exceptionally severe inflation. By 1973, prices were rising at rates around four times as high as during the 1960s. Real income was declining, unemployment was rising, and economic growth had come to a half. These factors clearly had an impact on the attitudes of Western publics. In December of each year, a national sample of the German public is asked, "Is it with hopes or with fears that you enter the New Year?" In December 1968, optimism was at an all-time high: 65% said that they felt hopeful. Confidence remained almost equally high in December 1969, just before our first survey: 63% felt hopeful. There was a marked deterioration of confidence in subsequent years. By December 1973 (just after our most recent survey) only 30% of those sampled expressed hope; confidence among the German public was at its lowest point since 1950. American consumer confidence reached its lowest recorded level in early 1974.[7] And in 1972 and 1973, heavy pluralities of the publics of Germany, France, Italy, Belgium, and the Netherlands expressed the opinion that the economic situation had deteriorated during the previous year. In January 1974, 84% of the Italian public felt that the economic situation had become worse during the previous year, as compared with 5% who felt it had improved. The outlook was almost equally gloomy in the Netherlands, Germany, and Belgium; only the French remained even relatively optimistic —which is to say that negative perceptions predominated over positive ones by nearly two to one (Commission of European Communities, 1974).

If such conditions persisted for very long, we would expect them to reshape the priorities of Western publics in an increasingly materialist direction. But the question is, "*How* long?" If the items that we have been discussing really do tap an individual's basic value priorities, they should normally be reasonably resistant to short-term forces.

Let us examine changes in the distribution of value types in the six countries for which we have data at both points in time. Table 5 shows the comparison between 1970 and 1973. There was little indication of an increase in the proportion of materialists in these six

Table 5. CHANGES OVER TIME IN DISTRIBUTION OF VALUE TYPES*
(February-March 1970 versus September-October 1973)

	1970	1973
Germany		
Materialist	43%	42%
Postmaterialist	10	8
France		
Materialist	38	35
Postmaterialist	11	12
Italy		
Materialist	35	40
Postmaterialist	13	9
Belgium		
Materialist	32	25
Postmaterialist	14	11
Netherlands		
Materialist	30	31
Postmaterialist	17	13
Great Britain		
Materialist	36	32
Postmaterialist	8	8

*1970 figures are from "The Silent Revolution in Europe," *American Political Science Review* (December 1971), p. 995. The original 4-item index is used for both years.

countries. In most instances, the differences were so small that they could be attributed to normal sampling error. In several cases the proportion of materialists seems to have *declined,* not increased; and in one case (France), the percentage of postmaterialists showed a slight increase, rather than a decline. The overall change was in the direction that we would expect under conditions of economic decline, but when we compare it with the precipitous drop in consumer confidence registered for this period, it seems almost incredibly small.

The four-item materialist-postmaterialist indicator was first used with a national sample of the American public in May 1972; it was replicated in two other American surveys within less than a year.[8] All these surveys took place before the Arab oil embargo of October 1973; they showed a mean of 32% materialist and 11% postmaterialist among the American public. Our 1974 survey took place during a period of severe economic contraction—but it showed a *lower* proportion of materialists and a *higher* proportion of postmaterialists than any previous survey (the figures were 27% and 14% respectively). Again, the shifts were small enough to be attributed to sampling error—but they moved *counter* to the trend that one would expect if response to these items simply reflected current events.

An examination of times series data based on response to the Rokeach Terminal Values survey gives a similar impression of stability in the face of short-term forces. Rokeach administered this battery to national samples of the American public in 1968 and 1971; our 1974 replication provides data covering a six-year span. The aggregate rankings of his 18 items are shown in Table 6.

The results show a truly remarkable degree of continuity. Among the six items ranked highest in 1968, not one fell lower than seventh place at either subsequent time point. Similarly, the five items ranked lowest in 1968 remained the bottom five in both 1971 and 1974, and most of them budged no more than one rank above or below their original position. There was less stability among the items given an intermediate ranking—apparently people can single out most clearly and reliably the things that seem either terribly important or quite unimportant. We should note that "national security" fell within this intermediate range. However, even these items invariably remained within the middle zone, never rising to a high rank or dropping to a very low one. It seems fair to conclude that the battery as a whole tapped something deep-rooted in the outlook of the American public.

Table 6. STABILITY OF PERSONAL VALUES AMONG THE AMERICAN PUBLIC, 1968-1974 (Rankings of Rokeach's Terminal Values)*

	Composite Ranking		
Values	**1968**	**1971**	**1974**
A WORLD AT PEACE (free of war and conflict)	1	1	2
FAMILY SECURITY (taking care of loved ones)	2	2	1
FREEDOM (independence, free choice)	3	3	3
HAPPINESS (contentedness)	4	6	5
SELF-RESPECT (self-esteem)	5	5	4
WISDOM (a mature understanding of life)	6	7	6
EQUALITY (brotherhood, equal opportunity for all)	7	4	12
SALVATION (saved, eternal life)	8	9	10
A COMFORTABLE LIFE (a prosperous life)	9	13	8
A SENSE OF ACCOMPLISHMENT (lasting contribution)	10	11	7
TRUE FRIENDSHIP (close companionship)	11	10	9
NATIONAL SECURITY (protection from attack)	12	8	13
INNER HARMONY (freedom from inner conflict)	13	12	11
MATURE LOVE (sexual and spiritual intimacy)	14	14	14
A WORLD OF BEAUTY (beauty of nature and the arts)	15	15	15
SOCIAL RECOGNITION (respect, admiration)	16	17	18
PLEASURE (an enjoyable, leisurely life)	17	16	16
AN EXCITING LIFE (a stimulating, active life)	18	18	17

*SOURCE: National probability samples interviewed by NORC in spring 1968 and spring 1971 and by Survey Research Center in summer 1974; respective Ns: 1,409; 1,430; 1,719.

But even granting that our values items do tap something basic, perhaps linked with the formative experiences of given generations, we are not in a position to make inferences about long-term attitudinal change—for we have not yet established that there is any relationship between one's basic values and one's attitudes toward military service. A priori, it would certainly seem reasonable to expect that those for whom national security has a relatively high priority would tend to be supportive of military service. But a voluminous literature demonstrates that one often finds amazingly little relationship or "constraint" among the attitudes of mass publics. It would be rash to take such a linkage for granted.

Attitudes Toward Military Service

The 1974 survey contained a set of questions that should enable us to explore the relationship between our values indicators and attitudes toward military service. These questions and the public's responses to them appear in Table 7.

Perhaps the most striking aspect of Table 7 is the extent to which the general public seems to have a positive image of military service. The only item in Table 7 which does not evoke a clearly favorable response is "Life in the armed forces is not as interesting as the ads say," and this finding may reflect skepticism about whether *anything* is as attractive as the ads say, more than a negative feeling toward military service. This favorable view among the general public hardly corresponds to the impression that one is likely to have gained from the perspective of college campuses, but it is consistent with another piece of evidence about the broader public that is available from our survey. Our respondents were asked to rate "the military" and other American institutions on a feeling thermometer ranging from zero (the negative pole) to 100 (the most favorable possible rating). The military was given a positive rating by the vast majority of the public and, as Table 8 demonstrates, its aggregate rating compares quite favorably with that of other groups. This may seem surprising, in the light of the unfavorable publicity which the military has received in recent years, but it confirms other recent findings (Rodgers and Johnston, 1974; D. Segal, 1975a).

At first glance, these results might seem somewhat paradoxical. The American public does not accord a particularly high priority to

Table 7. ATTITUDES TOWARD MILITARY SERVICE AMONG THE AMERICAN PUBLIC, SUMMER 1974

Question / Response	%
1. If you had a son in his late teens or early twenties who decided to enter the military service, how would you feel about his decision?	
Strongly positive	30%
Mostly positive	47
Mostly negative	12
Strongly negative	5
2. Would you say that this country's armed forces today are absolutely essential, fairly essential, or not at all essential?	
Absolutely essential	68%
Fairly essential	28
Not at all essential	2
3. What sort of career do you think the armed forces offer today?	
A very good career	40%
Quite a good career	44
Not a very good career	9
Do you agree or disagree with the following statements that have been made about life in the armed forces?	
4. The pay is good.	
Agree	63%
Disagree	19
5. Life in the armed forces is not as interesting as the ads say.	
Agree	48%
Disagree	34
6. Life in the armed forces is more like civilian life than it used to be.	
Agree	77%
Disagree	9
7. The trouble with the armed forces is that you have to sign up for such a long period of time.	
Agree	32%
Disagree	54
8. Life in the armed forces is unsuitable for a man or woman with a family.	
Agree	37%
Disagree	53
9. Have you ever served in any branch of the military service?	
Yes	20%
No	79

national defense, yet it holds a clearly positive view of the military and military service. However, this pattern implies that people tend to place a positive value on most or all of the goals included in our pool of 30 items. But the findings indicate that there is no one-to-one relationship between value priorities and support for military service; our reluctance to assume that there would be seems well founded.

Table 8. RATINGS OF GROUPS ON THE "FEELING THERMOMETER"
(Mean rating by cross-section of the American public, 1974)

Clergymen (priests, ministers, rabbis)	80	
Small Businessmen	78	
The Police	78	
Whites	75	
The Military	73	
Civil Service Employees	67	
Blacks	63	
Migrants Workers	61	
The Democratic Party	61	
Labor Unions	57	Positive
Big Business	55	↑
The Republican Party	50	——
The Women's Liberation Movement	48	↓
Student Protesters	32	
Revolutionary Groups	21	Negative

We will not attempt to analyze each of the items in Table 7 separately. They were intended to measure support for military service; factor analysis indicates that they do tap this underlying dimension–though with varying degrees of success. The item concerning one's feelings about a hypothetical son's decision to enter military service (which shows a loading of .628) is our most sensitive indicator of this underlying attitude.[9] More specific items about the pay, the length of service required, and compatibility with family life (with loadings of .359, .356, .299 respectively) are weaker indicators; and one's perception of whether military life has become more like civilian life has relatively little to do with one's global assessment: the faint positive loading (.168) which this item shows might well be attributable to response set. On the other hand, one's rating of "the military" on the feeling thermometer *does* have a strong correlation with attitudes toward a military career. Though they go together empirically, there are some interesting differences between the pattern of responses to the question about a son's military service and ratings of the military on the feeling thermometer. The former seems more likely to evoke immediate practical considerations, while the latter measures one's feelings of basic sympathy or distaste for the military–and thus probably taps a more deep-rooted attitude than the former. Both elements enter into one's assessment of military service, but the former is subject to relatively pronounced changes over short periods of time.

To illustrate this point, let us turn to Table 9. The data in this

Table 9. ATTITUDES TOWARD MILITARY SERVICE DURING AND AFTER U.S. MILITARY INVOLVEMENT IN VIETNAM

"If you had a son in his late teens who decided to enter military service, how would you feel about his decision?"

	Late 1972-Early 1973	Summer 1974
Strongly positive	19%	30%
Mostly positive	41	47
Mostly negative	19	12
Strongly negative	8	5

table were gathered in two surveys carried out about a year and a half apart. There was a substantial increase in the proportion of favorable attitudes toward military service during this relatively brief interval. It is not difficult to think of reasons why this sudden change occurred. American forces were still in Vietnam during the former period; they had withdrawn by the later time point. Furthermore, economic conditions had deteriorated by the time of the 1974 survey to such an extent that a military career might have seemed somewhat more attractive than it did earlier. The point is that a sizable change took place in response to this item.

Thermometer ratings of the military showed considerably more stability. As Table 10 indicates, these ratings varied only marginally during the entire decade from 1964 to 1974—from a time when American involvement in Vietnam still seemed very minor, through the peak of American involvement, to a time when American forces had been withdrawn. There was indeed a decline in affect for the military when American involvement was at its height, but the size of the maximum fluctuation across the 10-year period was smaller than the changes during the 18-month period shown in Table 9. The relative stability of affect ratings does not seem to be an artifact

Table 10. RATINGS OF VARIOUS GROUPS ON THE "FEELING THERMOMETER," 1964-1974 (Mean rating by national cross-sections)

	1964	1968	1972	1974	Change 1964-1974
Whites	83	80	77	75	− 8
The Military	75	74	70	73	− 2
Democrats/the Democratic Party	71	60	60	61	−10
Negroes/Blacks	63	64	64	63	0
Big Business	60	59	53	55	− 5
Republicans/the Republican Party	59	62	63	50	− 9
Labor Unions	58	56	56	57	− 1

owing to the type of scale which was used to measure them: the ratings of certain other groups showed substantial changes during this period.

We suggest that one reason why overall affect toward the military was more stable than feelings about a son's military career is that the former is more closely linked with one's basic values. For these attitudes are, in fact, correlated with value priorities, as we hypothesized; and the correlation between values and overall affect toward the military was about twice as strong as the correlation with feelings about a son's military career. Since the postmaterialists tended to be younger and better educated than the materialists, it is not surprising to find that this pattern of differential correlations also applied to the relationships with age and education.

Table 11 illustrates the relationship between age and the two respective attitudes. In both cases, the young were less likely to have favorable feelings toward military service and the military than were the old; but the relationship with overall feelings toward the military was clearly stronger. Similarly, the better educated were less likely to have favorable feelings than were the less educated, as Table 12 shows. But the relationship with overall affect was the stronger of the two by a wide margin.

The younger and the better educated were less favorable toward military service than the older and less educated: the pattern replicates earlier findings reported by Jerald G. Bachman, in an exhaustive investigation of the subject (1973, 1974).[10] And we concur also with Bachman's finding that age and education are the most important of the standard social background predictors of these attitudes; other cleavages seem relatively minor. Our interpretation is that age and education show these relatively strong linkages because

Table 11. ATTITUDES TOWARD MILITARY SERVICE AND "THE MILITARY" BY AGE, AMONG THE AMERICAN PUBLIC, 1974

Ages	Percentage feeling "strongly positive" about a son's military service		Percentage rating "the military" above 75 on the feeling thermometer	
18-24	23%	(313)	36%	(313)
25-34	24	(360)	42	(359)
35-44	30	(268)	49	(268)
45-54	36	(266)	59	(262)
55-64	34	(236)	60	(229)
65+	35	(276)	64	(261)
Gamma	.101		.228	

Table 12. ATTITUDES TOWARD MILITARY SERVICE AND "THE MILITARY" BY EDUCATION, AMONG THE AMERICAN PUBLIC, 1974

Educational level	Percentage feeling "strongly positive" about a son's military service		Percentage rating "the military" above 75 on the feeling thermometer	
Less than 12th grade	35%	(627)	64%	(610)
High school graduate	28	(531)	51	(527)
Beyond high school	25	(561)	36	(555)
Gamma	.078		.310	

they are distinctive social characteristics of the postmaterialists —those raised in relative economic and physical security. But we must ask, just how strong is the relationship between values and attitudes toward military service? Does it really have an impact on these attitudes, or is the relationship merely due to mutual linkages with age and education? Let us try to answer each of these questions.

Our response to the first question is, "It depends." For survey research must normally contend with considerable error in measurement, and one's value priorities are particularly difficult to measure; the strength of the relationship depends on the quality of one's measure. We spent a good deal of time in the early part of this paper discussing various measures of value priorities, gradually moving from our original four-item measure to more broadly based and presumably more accurate indicators of the materialist-postmaterialist dimension. Table 13 suggests that the effort was worthwhile, for the correlation between values and our dependent variable rises quite markedly as we move from our earliest and narrowest measure to one based on the much larger pool of items available in the 1974 survey.

Table 13. RELATIONSHIP BETWEEN SUPPORT FOR MILITARY SERVICE AND THREE MEASURES OF VALUE PRIORITIES: U.S., 1974*
(Product-moment correlations)

		r =
	Original 4-item values index	.149
Factor scores, first dimension	12 Materialist-Postmaterialist items	.259
	18 items (11 Materialist-Postmaterialist items plus 7 related Rokeach items)	.329

*Dependent variable is measured by factor scores generated by the analysis shown in Table 9.

Table 14. SUPPORT FOR MILITARY SERVICE BY VALUE TYPE (Percentage whose factor scores reflect relatively strong support for military service)*

Value Type (collapsed factor scores)**		
Postmaterialist	0%	(16)
Score = 2	15	(40)
Score = 3	16	(61)
Score = 4	39	(113)
Score = 5	43	(175)
Score = 6	46	(291)
Score = 7	49	(156)
Score = 8	55	(309)
Score = 9	59	(232)
Materialist	56	(78)

*Factor scores generated by the analysis of the items in Table 6 provide the dependent variable; they were dichotomized at a point where 47% of the total sample ranks "high."
**Value type is based on factor scores generated by the first dimension in an analysis of the 11 materialist-postmaterialist items plus the 7 related Rokeach Terminal Values indicated in Table 4.

Table 14 translates the latter correlation coefficient into an array of percentages, giving a more concrete picture of the effect. As we move from the postmaterialist to the materialist end of the continuum, there is a pronounced and fairly steady rise in support for military service.

Our second question is more difficult to answer. One can rarely demonstrate a causal relationship, particularly in the absence of time-series data. But we can inquire whether the observed relationship between values and attitudes can be explained, in a statistical sense, by the linkages with age and education. The answer seems to be "no." Table 15 shows the results of a multiple regression analysis involving the three key predictors of support for military service. Applying statistical controls for the effects of age and education

Table 15. VALUE PRIORITIES AND SUPPORT FOR MILITARY SERVICE, CONTROLLING FOR AGE AND EDUCATION (Multiple regression analysis)

Variable	Partial Correlation
Value priorities*	.280
Age	–.134
Years of education	.013
Multiple correlation =	.356

*Our measure of value priorities is the factor score generated by the 18 items indicated in Table 4. The dependent variable is the factor score generated by the dimension shown in Table 9. Number of respondents was 1,461.

somewhat weakens the correlation between values and the dependent variable but by no means causes it to disappear. Values explain more variance than the other two variables combined. Postmaterialists do, indeed, tend to be relatively young—but their distinctive attitudes toward military service seem to reflect their value priorities, rather than anything inherent in age (or education) in themselves. We would tentatively conclude that one's value priorities have a significant impact on attitudes toward military service.

Conclusions

If the value priorities of the American people are indeed changing, this would constitute a force tending gradually but steadily to undermine public support for military institutions. The American public has, at present, a predominately positive image of military service and a generally sympathetic view of the military. Moreover, attitudinal changes in the last few years seem to have been favorable. Furthermore, although cross-sectional data suggest that values have been changing in a postmaterialist direction during the last few decades, there is nothing inevitable about this trend: if (as we hypothesize) it has resulted from the presence of relatively high levels of economic and physical security, the termination of these conditions should gradually reverse the trend. If unfavorable economic conditions were to persist for many years, or if the United States were plunged into a major war, one would expect a decline in the proportion of postmaterialists.

Assuming, on the other hand, that there will be a revival of prosperity and continued peace, the process of value change would apparently continue. This would not in itself ensure an automatic decline in mass support for military institutions. In a given situation, different motivations can lead to similar attitudes (Schuman, 1972). One can conceive other circumstances under which both materialists and postmaterialists would wholeheartedly support the maintenance of strong defense forces.

This would probably require a rather extraordinary set of circumstances. It seems likely that our 1974 data reflect the relationship that one would normally find. If this is true, we may be witnessing a gradual but protracted erosion of public support for military institutions.

Notes

1. For details of how this values typology was constructed, a description of the data base, and a more complete presentation of our hypotheses, see Inglehart, 1976.

2. In the American survey, the item "protecting nature from being spoiled and polluted" was used instead of "trying to make our cities and countryside more beautiful." Neither of these items was particularly effective in tapping the intended dimension.

3. For this analysis, each item was recoded as a separate variable with codes ranging from 1 to 6. If the given item was chosen as the "most desirable" among the entire set of items, it was coded as 1; if it ranked second overall, it was coded as 2; if it ranked last overall it was coded as 6. If chosen first among its set of four items (but not first or second overall) it was coded 3; if ranked second in its set of four, it was coded as 4; items not singled out for either high or low rankings were coded as 5.

4. The 1974 American survey was carried out by the fieldwork staff of the Institute for Social Research at the University of Michigan, from July 1974 to September 1974; S.H. Barnes and M.K. Jennings were the coinvestigators. A representative national cross-section of Americans aged 18 and older was interviewed (N = 1,719), together with a supplementary sample of 245 respondents integrated with our main sample in such a way as to provide 245 parent-child pairs, for intergenerational comparisons. The present paper reports percentaged responses from the national cross-section only.

5. In 1974 the original four-item set was kept apart from the eight newer items: each respondent indicated his first, second, and third highest-rankings within the four-item set; and then he gave a separate set of rankings (the top three, and the bottom three choices) within the eight-item set. Thus, any spurious constraint that might have existed between the two sets of items owing to the fact that they were ranked in a common pool in 1973 would be absent in 1974.

6. Rokeach uses two sets of items, consisting of 18 Instrumental Values and 18 Terminal Values. We made use of the Terminal Values battery only.

7. Similarly, in December 1973, the Sindlinger Consumer Confidence Index showed the most pessimistic outlook in more than 25 years.

8. The first of these surveys was carried out by the Survey Research Center of the Institute for Social Reseach, University of Michigan. This measure was replicated as part of the 1972 election survey by the Center for Political Studies (also of the Institute for Social Research). The third survey was sponsored by the European Community's Washington office, with fieldwork carried out by the Gallup organization in March 1973.

9. This item was replicated from a survey carried out by Jerald G. Bachman (1973, 1974). Data from his survey appear in Table 10. Bachman's research indicated that this was a particularly effective item for measuring one's attitude toward military service in general.

10. Still earlier, Rokeach (1973:64, 76) noted that the younger and better educated gave a relatively low priority to "national security."

BIBLIOGRAPHIC REFERENCES

BIBLIOGRAPHIC REFERENCES

ABRAHAMSSON, B. (1970). "Elements of military conservatism." Paper read at the seventh World Congress of Sociology, Varna, Bulgaria, August.

ALLEN, H.E. (1972). "Schilling Manor: A survey of a military community of father absent families." Unpublished doctoral dissertation, Catholic University of America.

ARTHUR, R.J. (1966). "Psychiatric disorders in naval personnel." Military Medicine, 131:354-361.

--- (1971). "Success is predictable." Military Medicine, 136:539-545.

BACHMAN, J.G. (1970). Youth in transition: The impact of family background and intelligence on tenth grade boys. Ann Arbor: Institute for Social Research, University of Michigan.

--- (1973). Values, preferences and perceptions concerning military service (pt. 1). Ann Arbor: Institute for Social Research, University of Michigan.

--- (1974). Values, preferences and perceptions concerning military service (pt. 2). Ann Arbor: Institute for Social Research, University of Michigan.

BACHMAN, J.G., GREEN, S., and WIRTANEN, I.D. (1971). Youth in transition: Dropping out–problem or symptom? Ann Arbor: Institute for Social Research, University of Michigan.

BACHMAN, J.G., and JENNINGS, M.K. (1975). "The impact of Vietnam on trust in government." Journal of Social Issues, 31(4):1410156.

BACHMAN, J.G., and JOHNSTON, J. (1972). "The all-volunteer force: Not whether but what kind?" Psychology Today, 5(October):113-116, 128.

BACHMAN, J.G., KAHN, R.L., MEDNICK, M.T., DAVIDSON, T.N., and JOHNSTON, L.D. (1967). Youth in transition: Blueprint for a longitudinal study of adolescent boys. Ann Arbor: Institute for Social Research, University of Michigan.

BACK, K.W. (1973). "Neglected psychological issues in population research." American Psychologist, (July):567-572.

BAIN, E.H. (1972). "Personality leadership, and motivational characteristics of ROTP resignees from RMC" (Research Report 72-8 Department of MLM, RMC). Kingston, Ont.

--- (1973). "The socialization process of ROTP cadets at RMC: A study of personality, leadership, values, and motivational characteristics of stayers and leavers" (Research Report 73-3). Kingston, Ont.: Department of Military Leadership and Management, Royal Military College.

BAKER, S. (1967). "Families under stress during dads' absence." United States Medicine, pp. 6-7.

BAKER, S., COVE, L., FAGEN, S., FISCHER, E., and JANDA, E. (1968). "Impact of father absence: III. Problems of family reintegration following prolonged father absence" (abstract). American Journal of Orthopsychiatry, 38:347.

BAKER, S., FAGEN, S., FISCHER, E., JANDA, E., and COVE, L. (1967). "Impact of father absence on personality factors of boys: I. An evaluation of the military family's adjustment." American Journal of Orthopsychiatry, 37:269.

BARBER, J.A., Jr. (1972). "The social effects of military service." Pp. 151-166 in S.E. Ambrose and J.A. Barber, Jr. (eds.), The military and American society. New York: Free Press.

BARKER, R.G. (1968). Ecological psychology. Stanford, Calif.: Stanford University Press.

BARKER, R.G., and GUMP, P.V. (1964). Big school, small school. Stanford, Calif.: Stanford University Press.

BARNES, P.P. (1972). The plight of the citizen soldier. New York: Knopf.

BECKE, M. (1973). "A comparative study of the level of social maturity of military dependent and nonmilitary dependent adolescents." Unpublished masters thesis, Stetson University.

BELLINO, R. (1969). "Psychomatic problems of military retirement." Psychosomatics, 10:318-321.

--- (1970). "Perspectives of military and civilian retirement." Mental Hygiene, 54:580-583.

BELT, J., and SWENEV, A. (1973). "The Air Force wife: Her knowledge of, and attitudes toward, the Air Force." Paper presented at the Military Testing Association Conference on Human Resources, San Antonio, Texas, October.

BENNETT, L. (1944). "Problems of homecoming." Survey Midmonthly, pp. 246-248.

BENNETT, W.M., Jr., CHANDLER, H.T., DUFFY, J.P., Jr., HICKMAN, J.L., JOHNSON, C.R., LALLY, M.J., Jr., NICHOLSON, A.L., NORBO, G.J., OMPS, A.R., POSPISIL, V.A., SEEBERG, R.S., and WUBBENA, W.L., Jr. (1974). Army families. Carlisle Barracks: U.S. Army War College Military Research Program.

BEVILACQUA, J. (1967). "Civilianization and health-welfare resource participation on an Army post." Unpublished doctoral dissertation, Brandeis University.

BEY, D.R., Jr., and LANGE, J. (1974). "Waiting wives: Women under stress." American Journal of Psychiatry, 131:283-286.

BEY, D.R., Jr., and SMITH, W.E. (1971). "Organizational consultation in a combat unit." American Journal of Psychiatry, 128(October):401-406.

BIDERMAN, A. (1959). "The prospective impact of large scale military retirement." Social Problems, 17:84-90.

--- (1964). "Sequels to a military career: The retired military professional." Pp. 287-338 in M. Janowitz (ed.), The new military: Changing patterns of organization. New York: Russell Sage Foundation.

--- (1971). "The retired military." Pp. 123-163 in R. Little (ed.), Handbook of military institutions. Beverly Hills, Calif.: Sage.

--- (1972). "Retired soldiers within and without the military-industrial complex." Pp. 95-126 in S. Sarkesian (ed.), The military-industrial complex: A reassessment. Beverly Hills, Calif.: Sage.

BIDERMAN, A., and SHARP, L.M. (1968). "The convergence of military and civilian occupational structures." American Journal of Sociology, 73(January):381-399.

BINKIN, M., and JOHNSTON, J.D. (1973). All-volunteer armed forces: Progress, problems, and prospects. Report prepared for the Committee on Armed Services, U.S. Senate, 93rd Congress, 1st Session. Washington, D.C.: U.S. Government Printing Office.

BITTNER, E. (1967). "The police on skid-row: A study of peacekeeping." American Sociological Review, 32(October).

BLAIR, J.D. (1975). "Civil-military belief systems: A study of attitudes toward the military system among military men and civilians." Unpublished Ph.D. dissertation, University of Michigan.

BLETZ, D. (1972). The role of the military professional in U.S. foreign policy. New York: Praeger.

BLOCH, H.S. (1969). "Army psychiatry in the combat zone–1967-1968." American Journal of Psychiatry, 126:291-292.

BLOCHBERGER, C.W., Jr. (1970). "Military families: Differential lifestyles." Unpublished doctoral dissertation, University of California, Berkeley.

BLONSTON, G. (1974). "ROTC goes co-ed–and grows." Philadelphia Inquirer, October 20.

BLOOM, B.S. (1964). Stability and change in human characteristics. New York: John Wiley.

BOOTH, R.F., and HOIBERG, A. (1973). "Change in Marine recruits' attitudes related to recruit characteristics and drill instructors' attitudes." Psychological Reports, 33:63-71.

--- (1974). "Structure and measurement of Marine recruit attitudes." Journal of Applied Psychology, 59:236-238.

BORUS, J.F. (1973a). "Reentry: I. Adjustment issues facing the Vietnam returnee." Archives of General Psychiatry, 28:501-506.

--- (1973b). "Reentry: II. 'Making it' back in the States." American Journal of Psychiatry, 130:850-854.

--- (1973c). "Reentry. III. Facilitating healthy readjustment in Vietnam veterans." Psychiatry, 36:428-439.

--- (1974). "Incidence of maladjustment in Vietnam returnees." Archives of General Psychiatry, 30:554-557.

BORUS, J.F., FINMAN, B.G., and STANTON, M.D., and DOWD, A.F. (1973). "The racial perceptions inventory." Archives of General Psychiatry, 29:270-275.

BORUS, J.F., STANTON, M.D., FINMAN, B.G., and DOWD, A.F. (1972). "Racial perceptions in the Army: An approach." American Journal of Psychiatry, 128:1369-1374.

BOULDING, E. (1950). "Family adjustments to war separations and reunion." Annals of the American Academy, pp. 59-67.

BOWER, E. (1967). "American children and families in overseas communities." American Journal of Orthopsychiatry, 37:787-796.

BOWERS, D.G., and BACHMAN, J.G. (1974). Military manpower and modern manpower. Ann Arbor: Institute for Social Research, University of Michigan.

BOWERS, D.G., and FRANKLIN, J.L. (1973). The Navy as a functioning organization: A diagnosis. Ann Arbor: Institute for Social Research, University of Michigan.

BRAATZ, G.A., LUMRY, G.K., and WRIGHT, M.S. (1971). "The young veteran as a psychiatric patient in three eras of conflict." Military Medicine, 136:455-457.

BRADY, D., and RAPPOPORT, L. (1973). "Violence and Vietnam." Human Relations, 26:735-752.

BRONFENBRENNER, U. (1974). "The origins of alienation." Scientific American, 231:53-61.

BROWN, D., CATALDO, J., DAVISON, R., and HUYCKE, E. (1973). "Problems of Vietnam prisoners of war and their families." Social Psychiatry, 1.

BROWN, M. (1944). "When our servicemen come home." Journal of Home Economics, pp. 626-628.

BROWNING, H., LOPREATO, S.C., and POSTON, D.L., Jr. (1973). "Income and veteran status." American Sociological Review, 38(February):74-85.

BURCHARD, W.W. (1954). "Role conflicts of military chaplains." American Sociological Review, 19(October):528-535.

BUTLER, R.P. (n.d.). "Aptitude for the service rating as a predictor of junior officer performance." West Point, N.Y.: U.S. Military Academy.

--- (1974). "Prediction of officer performance." West Point, N.Y.: U.S. Military Academy.

CABANILLAS, C.E. (1975). "The Army officer's wife: Role conflict and role strain." Unpublished M.A. thesis, University of Texas, El Paso.

CAMPBELL, D.T., and McCORMACK, T.H. (1957). "Military experience and attitudes toward authority." American Journal of Sociology, 62:482-490.

CAMPBELL, J. (1956). The hero with a thousand faces. New York: Meridian.

CANTRELL, R.W. (1974). "Prolonged exposure to intermittent noise: Audiometric, biochemical, motor, psychological and sleep effects." The Laryngoscope, 84(suppl. 1).

CAPLAN, G. (1971). "Testimony on unemployment and overall adjustment problems of returning veterans." Pp. 38-58 in Hearings before the Subcommittee on Veterans' Affairs of the Committee on Labor and Public Welfare, U.S. Senate. Washington, D.C.: U.S. Government Printing Office.

––– (1975). "Organization of support systems for civilian populations." Paper presented at the International Conference on Psychological Stress and Adjustment in Time of War and Peace, Tel Aviv, Israel, January.

CARPENTER, G.J. (1973). "Cadet retention at the Royal Military College of Canada as a function of personality leadership performance, and biographical variables" (Research report 73-4). Kingston, Ont.: Department of Military Leadership and Management, Royal Military College.

CARROLL, R. (1974). "Ethics of the military profession." Air University Review, (November-December):42.

CHAPMAN, W. (1974). "Military most admired U.S. institution." Washington Post, May 9.

CHRISTIE, R. (1952). "Changes in authoritarianism as related to situational factors." American Psychology, 7:307-308.

––– (1954). Transition from civilian to Army life (HumRRO Technical report 54-13). Alexandria, Va.: Human Resources Research Office.

CLINE, V. (1955). "A survey of opinions regarding operation gyroscope in the first division" (staff memorandum). Washington, D.C.: Human Resources Research Office, George Washington University.

CLUM, G.A., and HOIBERG, A.L. (1971). "Prognostic indexes in a military psychiatric population." Journal of Consulting and Clinical Psychology, 36:436-440.

COATES, C.H., and PELLEGRIN, R.J. (1965). Military sociology: A study of American military institutions and military life. Baltimore: Social Science Press.

COLEMAN, J.S. (1961). The adolescent society. Glencoe, Ill.: Free Press.

Commission of the European Communities (1974). "Information memo: Results of the sixth survey on consumers' views of the economic situation" (April). Brussels.

CONVERSE, P.E. (1969). "Of time and partisan stability." Comparative Political Studies, 2(July):139-171.

CUBER, J.F. (1945). "Family readjustment of veterans." Marriage and Family Living, pp. 28-30.

CUTRIGHT, P. (1974). "The civilian earnings of white and black draftees and nonveterans." American Sociological Review, 39(June):317-327.

DAHL, B., and McCUBBIN, H. (1974). "The adjustment of children of returned prisoners of war: A preliminary report." Paper presented at the Second Joint Medical Meeting on Prisoners of War, Department of Defense, San Antonio, Texas, November.

––– (1975). "Children of returned prisoners of war: The effects of long-term father absence." Paper presented at the International Conference on Dimensions of Anxiety and Stress, Oslo, Norway, June.

DANIELS, A.K., and DANIELS, R.R. (1964). "The social function of the career fool." Psychiatry, 27:218-229.

DARNAUER, P.F. (1969). "Army brats–Growing up in an Army family." Proceedings from the ACS Workshop: 1969. Washington, D.C.: U.S. Department of the Army, Office of the Deputy Chief of Staff for Personnel.

DENTLER, R.A., and ERIKSON, K.T. (1959). "The functions of deviance in groups." Social Problems, 7(fall):98-107.

De SHALIT, N. (1970). "Children in war." In A. Jarus and J. Marcus (eds.), Children and families in Israel. New York: Gordon and Breach.

DICKERSON, W.J., and ARTHUR, R.J. (1965). "Navy families in distress." Military Medicine, 130:894-898.

DOLL, R.E., RUBIN, R.T., and GUNDERSON, E.K.E. (1969). "Life stress and illness patterns in the U.S. Navy: II. Demographic variables and illness onset in an attack carrier's crew." Archives of Environmental Health, 19:748-752.

DOODEMAN, L. (ed., 1974). "Mili-wife questionnaire results: The comment page wives speak out on the military, pro and con: Moving, children, rank consciousness, housing, benefits, women's liberation." Ladycom, 6:22-52.

DORNBUSCH, S. (1955). "The military academy as an assimilating institution." Social Forces, 33(May):316-321.

DOR-SHAV, N.K. (1975). "On the long-range effects of concentration camp internment on Nazi victims and their children." Paper presented at the International Conference on Psychological Stress and Adjustment in Time of War and Peace, Tel Aviv, Israel, January.

DRUSS, R. (1965). "Problems associated with retirement from military service." Military Medicine, 130:382-385.

DUNCAN, A. (1969). "Vietnam war widows learn to live and love again." Family Weekly, pp. 6-7.

DUVALL, E. (1945). "Loneliness and the serviceman's wife." Marriage and Family Living, 7:77-81.

EDWARDS, D., and BERRY, N.H. (1974). "Psychiatric decisions: An actuarial study." Journal of Clinical Psychology, 30:153-159.

ELIOT, T. (1946). "War bereavements and their recovery." Marriage and Family Living, 8:1-6.

ELOUL, J. (1975). "Basic issues in group work with war widows." Paper presented at the International Conference on Psychological Stress and Adjustment in Time of War and Peace, Tel Aviv, Israel, January.

FAGEN, S., JANDA, E., BAKER, S., FISCHER, E., and COVE, L. (1967). "Impact of father absence in military families: II. Factors relating to success of coping with crisis." Proceedings of the Annual Meeting of the American Psychological Association, 2.

FEDERMAN, P.J., LAUTMAN, M.R., and SIEGEL, A.I. (1973). Factors involved in the adjustment of low aptitude personnel to the Navy and their use for predicting reenlistment. Wayne, Pa.: Applied Psychological Sciences.

FIMAN, B.G., BORUS, J.F., and STANTON, M.D. (1975). "Black-white and American-Vietnamese relations among soldiers in Vietnam." Journal of Social Issues, 31(4):39-48.

FINLAYSON, E. (1969). "A study of the wife of an Army officer: Her academic and career preparation, her current employment and volunteer services." Unpublished doctoral dissertation, George Washington University.

FISHER, A.H., Jr. (1971). "Army 'new standards' personnel: Effect of remedial literacy training on performance in military service" (HumRRO Technical Report 71-7). Alexandria, Va.: Human Resources Research Office.

FLEISHMAN, E.A. (1953). "Leadership climate, human relations training, and supervisory behavior." Personnel Psychology, 6:205-222.

FORD, H.S. (1941). What the citizen should know about the Army. New York: W.W. Norton.

FRANCES, A., and GALE, L. (1973). "Family structure and treatment in the military." Family Process, 12:171-178.

FRANKLIN, J.L. (1973). A path analytic approach to describing causal relationships among social-psychological variables in multi-level organization. Ann Arbor: Institute for Social Research, University of Michigan.

FRASER, G.M. (1970). The general danced at dawn. New York: Ballantine.

FRENCH, E.G., and ERNEST, R.R. (1955). "The relationship between authoritarianism and acceptance of military ideology." Journal of Personality, 24:181-191.

FRENCH, J.R.P., Jr., RIDGERS, W., and COBB, S. (1974). "Adjustment as person-environment fit." Pp. 316-333 in C.V. Coelho, D.A. Hamburg, and J.E. Adams (eds.), New York: Basic Books.

FREUD, A., and BURLINGHAM, D. (1943). War and children. New York: International University Press.

FRIEDMAN, M. (1967). "Why not a volunteer Army?" Pp. 200-207 in S. Tax (ed.), The draft: A handbook of facts and alternatives. Chicago: University of Chicago Press.

FROEHLKE, R.F. (1972). "General officer promotion criteria detailed." Commander's Digest, July 20.

GABOWER, G. (1960). "Behavior problems of children in Navy officers' families." Social Casework, 41:177-184.

GANOE, W.A. (1924). The history of the United States Army. New York: D. Appleton.

GARBER, D.L. (1971). "Retired soldiers in second careers: Self-addressed change reference group salience, and psychological well-being." Unpublished doctoral dissertation, University of Southern California.

GARD, R. (1973). "The military profession." Naval War College Review, (July-August):12.

--- (1974). "Applications of behavioral science in the military environment." Paper presented at the meeting of the American Psychological Association, New Orleans, September.

GARDNER, G., and SPENCER, H. (1944). "Reactions of children with fathers and brothers in the armed forces." American Journal of Orthopsychiatry, 14:36-43.

GAULT, W.B. (1971). "Some remarks on slaughter." American Journal of Psychiatry, 128(October):450-453.

GEORGE, A.L. (1971). "Primary groups, organization and military performance." Pp. 293-318 in R.W. Little (ed.), Handbook of military institutions. Beverly Hills, Calif.: Sage.

GIFFEN, M.B., and McNEIL, J.S. (1967). "Effect of military retirement on dependents." Archives of General Psychiatry, 17:717-722.

GINZBERG, D. (1959). The ineffective soldier: The lost divisions. New York: Columbia University Press.

GITTER, A.G. (1973). The military: Redeployment of resources. Paper presented at the meeting of the American Psychological Association, Montreal, Canada, August.

GLICKMAN, A.S. (1961). The career motivation survey: Overall attitude and reenlistment trends (Research Report 61-2, June). U.S. Naval Personnel Research Field Activity.

GLICKMAN, A.S., GOODSTADT, B.E., KORMAN, A.K., and ROMANCZUK, A.P. (1973). Research and development for Navy career motivation program in an all-volunteer condition: I. A cognitive map of career motivation. Washington, D.C.: American Institutes for Research.

GOLAN, N. (1975). "From wife to widow to woman." Paper presented at the International Conference on Psychological Stress and Adjustment in Time of War and Peace, Tel Aviv, Israel, January.

GOLDMAN, N.L. (1973). "The changing role of women in the armed forces." American Journal of Sociology, 78(January):892-911.

GOLDSMITH, W., and CRETEKOS, C. (1969). "Unhappy odysseys: Psychiatric hospitalizations among Vietnam returnees." Archives of General Psychiatry, 20:78-83.

GONZALEZ, V.R. (1970). Psychiatry and the Army brat. Springfield, Ill.: C. C. Thomas.

GOODSTADT, B.E., FREY, R.L., Jr., and GLICKMAN, A.S. (1975). Socialization process and the adjustment of military personnel to Army life. Washington, D.C.: American Institutes for Research.

GOODSTADT, B.E., FREY, R.L., Jr., GLICKMAN, A.S., and ROMANCZUK, A.P. (1974). Socialization processes and the adjustment of military personnel to Army life: A structural analysis of socialization during advanced individual training. Washington, D.C.: American Institutes for Research.

GREENBERG, H. (1966). "Psychiatric symptomatology in wives of military retirees." American Journal of Psychiatry, 123:485-490.

GRIFFITH, C.R. (1944). "The psychological adjustments of returned servicemen and their families." Marriage and Family Living, 6:65-67, 87.

GRINKER, R.R., and SPIEGEL, J.P. (1945). Men under stress. New York: McGraw-Hill.

GUNDERSON, E.K.E. (1963). "Biographical indicators of adaptation to naval service" (Report 63-19, November). San Diego, Calif.: Navy Medical Neuropsychiatric Research Unit.

--- (1965). "Body size, self-evaluation and military effectiveness." Journal of Personality and Social Psychology, 2:902-906.

--- (1971). "Epidemiology and prognosis of psychiatric disorders in the naval service." Pp. 170-210 in C.D. Spielberger (ed.), Current topics in clinical and community psychology 3. New York: Academic Press.

GUNDERSON, E.K.E. and ARTHUR, R.J. (1966). "Demographic factors in the incidence of mental illness." Military Medicine, 131:429-433.

GUNDERSON, E.K.E., and JOHNSON, L.C. (1965). "Past experience, self-evaluation and present adjustment." Journal of Social Psychology, 66:311-321.

GUNDERSON, E.K.E., and RAHE, R.H. (1974). Life stress and illness. Springfield, Ill.: Charles C Thomas.

GUNDERSON, E.K.E., RAHE, R.H., and ARTHUR, R.J. (1970). "The epidemiology of illness in naval environments: II. Demographic, social background, and occupational factors." Military Medicine, 135:453-458.

GUNDERSON, E.K.E., RUSSELL, J.W., and NAIL, R.L. (1973). "A drug involvement scale for classification of drug abusers." Journal of Community Psychology, 1:399-403.

GUNDERSON, E.K.E., and SELLS, S.B. (1975). "Organizational and environmental factors in health and personnel effectiveness" (Report 75-8). San Diego, Calif.: Naval Health Research Center.

HAGGARD, E.A. (1949). "Psychological causes and results of stress." Pp. 441-461 in Human factors in undersea warfare. Washington, D.C.: National Research Council.

HALBERSTAM, D. (1969). The best and the brightest. Greenwich, Conn.: Fawcett Crest.

HALEY, J., and GLACK, I. (1965). Psychiatry and the family: an annotated bibliography of articles, 1960-1964. Palo Alto, Calif.: Family Process Monograph.

HALL, R., and SIMMONS, W. (1973). "The POW wife–A psychiatric appraisal." Archives of General Psychiatry, 29:600-694.

HALPERN, E. (1975). "Volunteering as a 'natural phenomena': An integration within Caplan's theory of support systems." Paper presented at the International Conference on Psychological Stress and Adjustment in Time of War and Peace, Tel Aviv, Israel, January.

HALPIN, A.W. (1954). "The leadership behavior and combat performance of airplane commanders." Journal of Applied Social Psychology, 49:19-22.

HAMBURG, D.A., and ADAMS, J.E. (1967). "A perspective of coping behavior." Archives of General Psychiatry, 17:277-284.

Harris, Louis, and Associates (1971). A study of the problems facing Vietnam era veterans: Their readjustment to civilian life. New York: Author.

--- (1973). Confidence and concern: Citizens view American government. Subcommittee on Intergovernmental Relations of Committee on Government Operations of the U.S. Senate. Washington, D.C.: U.S. Government Printing Office.

HARRIS, T.A. (1964). Counseling the serviceman and his family. New York: Prentice-Hall.

HARTNAGEL, T.F. (1974). "Absent without leave: A study of the military offender." Journal of Political and Military Sociology, 2:205-220.

HARTOG, J. (1966). "Group therapy with the psychotic and borderline military wives." American Journal of Psychiatry, 122:1125-1131.
HARWOOD, R. (1970). "Military under fire." Washington Post, July 12.
HAUSER, W.L. (1973). America's Army in crisis: A study in civil-military relations. Baltimore: Johns Hopkins University Press.
HELMER, J. (1974). Bringing the war home: The American soldier in Vietnam and after. New York: Free Press.
HOCHSCHILD, A.R. (1969). "The role of the ambassador's wife: An exploratory study. Journal of Marriage and the Family, 31(February):72-87.
HOIBERG, A., HYSHAM, C.E., and BERRY, N.H. (1974). "Effectiveness of recruits assigned to academic remedial training." Psychology Reports, 35:1007-1014.
HOLMES, T.H., and RAHE, R.H. (1967). "The social readjustment rating scale." Journal of Psychosomatic Research, 11:213.
HOLZ, R.F., and GITTER, A.G. (1974). Assessing the quality of life in the U.S. Army (Technical Paper 256). Arlington, Va.: U.S. Army Research Institute for the Behavioral and Social Sciences.
HUNT, W.A., and WITTSON, C.L. (1951). "The neuropsychiatric implications of illiteracy." U.S. Armed Forces Medical Journal, 2:365-369.
HUNTER, D., and PLAG, J. (1973). "An assessment of the needs of POW/MIA wives residing in the San Diego metropolitan area: A proposal for the establishment of family services" (Report 73-39). San Diego, Calif.: Navy Medical Neuropsychiatric Research Unit.
HUNTINGTON, S. (1974). "Postindustrial politics: How benign will it be?" Comparative Politics, 6(January).
IGEL, A. (1945). "The effect of war separation on father-child relations." Social Casework, 26:3-9.
INGLEHART, R. (1971). "The silent revolution in Europe: Intergenerational change in post-industrial society." American Political Science Review, 65(December):991-1017.
--- (1976). The silent revolution: Political change among Western publics. Princeton, N.J.: Princeton University Press.
INGLEHART, R., and BARNES, S.H. (1975). "System support and value change in the United States." Paper prepared for the Conference on Political Alienation, Iowa City, Iowa, January.
INSEL, P.M., and MOOS, R.H. (1974). "Psychological environments: Expanding the scope of human ecology." American Psychologist, (March):179-188.
ISAY, R. (1968). "The submariners' wives syndrome." Psychiatric Spectator, 5:18-19.
JANOWITZ, M. (1960). The professional soldier: A social and political portrait. New York: Free Press.
--- (1971a). "Basic education and youth socialization in the armed forces." Pp. 167-210 in R.W. Little (ed.), A handbook of military institutions. Beverly Hills, Calif.: Sage.
--- (1971b). The professional soldier: A social and political portrait (rev. ed.). New York: Free Press.
--- (1972a). "The decline of the mass Army." Military Review, (February):10-16.
--- (1972b). "Volunteer armed forces and military purpose." Foreign Affairs, (April): 428-443.
--- (1973). "The U.S. forces and the zero draft." Adelphi Papers, 94:1-30.
JANOWITZ, M., and LITTLE, R. (1965). Sociology and the military establishment (rev. ed.). New York: Russell Sage Foundation.
JANOWITZ, M., and MOSKOS, C., Jr. (1974). "Racial composition in the all-volunteer force: Policy alternatives." Armed Forces and Society, 1(fall):109-123.

JARUS, A., and MARCUS, J. (1971). Children and families in Israel. New York: Gordon and Breach.

JENNINGS, M.K., and MARKUS, G.B. (1974). "The effects of military service on political attitudes: A panel study." Paper presented at the meeting of the American Political Science Association, Chicago.

JENNINGS, M.K., and NIEMI, R.G. (1973). "Continuity and change in political orientations: A longitudinal study of two generations." Paper presented at the meeting of the American Political Science Association, New Orleans.

--- (1974). The political character of adolescence. Princeton, N.J.: Princeton University Press.

JOHNSON, L.C., SLYE, E., and LUBIN, A. (1965). "Autonomic response patterns during sleep." Paper presented at the meeting of the Association for Psychophysiological Study of Sleep, Bethesda, Md.

JOHNSON, V.W. (1967). Lady in arms. Boston: Houghton Mifflin.

JOHNSTON, J., and BACHMAN, J.G. (1972). Youth in transition: Young men and military service. Ann Arbor: Survey Research Center, Institute for Social Research, University of Michigan.

JOHNSTON, L. (1973). Drugs and American youth. Ann Arbor: Institute for Social Research. University of Michigan.

KAPLAN, R.C. (1962). "Personality correlates of military career choice." Unpublished Ph.D. thesis, University of Texas.

KARSTEN, P. (1972). The naval aristocracy. New York: Free Press.

KASSING, D.B. (1970). "Military experience as a determinant of veterans earnings." Pp. III-8-1 to III-8-22 in Studies prepared for the President's Commission on an All-Volunteer Armed Force. Washington, D.C.: U.S. Government Printing Office.

KATZ, D., GUTEK, B.A., and BARTON, E. (1974). Bureaucratic encounters: A pilot study in the evaluation of government service. Ann Arbor: Institute for Social Research, University of Michigan.

KEDEM, P. (1975). "The effect of the Yom Kippur War on the attitudes, values and locus of control of young adolescents." Paper presented at the International Conference on Psychological Stress and Adjustment in Time of War and Peace, Tel Aviv, Israel, January.

KELLER, S. (1963). Beyond the ruling class. New York: Random House.

KENNY, J. (1967). "The child in the military community." Journal of the American Academy of Child Psychiatry, 6:51-63.

KIMURA, Y. (1957). "War brides in Hawaii and their in-laws." American Journal of Sociology, 63:70-79.

KINZER, B., and LEACH, M. (1968). What every Army wife should know. Harrisburg, Pa.: Stackpole.

KIRSCHNER, E. (1975). "Pilot study on bereavement and rehabilitation of war widows of the Six-Day and Yom Kippur Wars." Paper presented at the International Conference on Psychological Stress and Adjustment in Time of War and Peace, Tel Aviv, Israel, January.

KOLB, D., GUNDERSON, E.K.E., and NAIL, R.L. (1974). "Pre-service drug abuse: Family and social history characteristics." Journal of Community Psychology, (July):278-282.

KOLB, D., NAIL, R.L., and GUNDERSON, E.K.E. (1975). "Pre-service drug abuse as a predictor of in-service drug abuse and military performance." Military Medicine, 140:104-107.

KORMAN, A.K. (1971). "Expectancies as determinants of performance." Journal of Applied Psychology, 55:218-222.

KOURVETARIS, G.A., and DOBRATZ, B. (1975). "The present state and development of military sociology." Paper presented at the American Sociological Association meeting, San Francisco, August.

KRISTAL, L. (1975). "The effects of a wartime environment upon the psychological development of children border settlements." Paper presented at the International Conference on Psychological Stress and Adjustment in Time of War and Peace, Tel Aviv, Israel, January.

KURLANDER, L., LEUKEL, D., PALEVSKY, L., and KOHN, F. (1961). "Migration: Some psychological effects on children." Paper presented at the annual meeting of the American Orthopsychiatric Association, New York.

LANG, K. (1965). "Military organizations." Pp. 838-878 in J. March (ed.), Handbook of organizations. Chicago: Rand McNally.

--- (1972). Military institutions and the sociology of war: A review of the literature with annotated bibliography. Beverly Hills, Calif.: Sage.

LaROCCO, J.M., and GUNDERSON, E.K.E. (1975). "Prediction of reenlistment: A discriminant analysis approach" (Report 75-21). San Diego, Calif.: Naval Health Research Center.

LaROCCO, J.M., GUNDERSON, E.K.E., DEAN, L.M., JAMES, L.R., JONES, A.P., and SELLS, S.B. (1975). "Organizational and environmental factors in health and personnel effectiveness: II. Data collection methods, test instruments, and criterion variables" (Report 75-9). San Diego, Calif.: Naval Health Research Center.

LASSWELL, H.D. (1941). "The garrison state." American Journal of Sociology, 46(January):455-468.

LAWRENCE, T.E. (1955). The mint: A day-book of the RAF depot between August and December 1922 with later notes by 352087 A/C Ross. Oxford: Alder.

LAZARUS, R.S. (1970). "Cognitive and personality factors underlying threat and coping." Pp. 143-164 in S. Levine and N.A. Scotch (eds.), Social stress. Chicago: Aldine.

LAZARUS, R.S., AVERILL, J.R., and OPTON, E.M., Jr. (1974). "The psychology of coping: Issues in research and assessment." Pp. 249-315 in G.V. Coelho et al. (eds.), Coping and adaptation. New York: Basic Books.

LEVI, L. (ed., 1971). Society, stress and disease: The psychosocial environment and psychosomatic diseases I. London: Oxford University Press.

LEVY, S. (1975). "Social work intervention in bereaved Israeli families." Paper presented at the International Conference on Psychological Stress and Adjustment in Time of War and Peace, Tel Aviv, Israel, January.

LIEBERMAN, E. (1971a). "American families and the Vietnam War." Journal of Marriage and the Family, 33:709-721.

--- (1971b). "War and the family: The psychology of antigrief." Modern Medicine, pp. 179-183.

LIFSCHITZ, M. (1975). "A follow-up study of bereaved elementary school children: Perceptual integration and adjustment." Paper presented at the International Conference on Psychological Stress and Adjustment in Time of War and Peace, Tel Aviv, Israel, January.

LIFTON, R.J. (1961). Thought reform and the psychology of totalism. New York: Norton.

--- (1968). Death in life: Survivors of Hiroshima. New York: Random House.

--- (1970a). "Introduction." History and human survival. New York: Random House.

--- (1970b). "On psychohistory." Partisan Review, (spring):11-32.

--- (1972). "Questions of guilt." Paritsan Review, 39(fall):514-530.

--- (1973a). Home from the war. New York: Simon and Schuster.

--- (1973b). "The sense of immortality: On death and the continuity of life." American Journal of Psychoanalysis, 33:3-15.

--- (1973c). "The struggle for cultural rebirth." Harpers, (April):84-90.

--- (n.d.). "The broken connection." Unpublished manuscript.

LINDQUIST, R. (1952). Marriage and family life of officers and airmen in a Strategic Air Command wing (Technical Report 5, Air Force Base Project). Chapel Hill: Institute for Research in Social Science, University of North Carolina.

LITTLE, R.W. (1971). "The military family." Pp. 247-270 in R.W. Little (ed.), Handbook of military institutions. Beverly Hills, Calif.: Sage.

LOPATA, H.Z. (1971). Occupation housewife. New York: Oxford University Press.

LOPEZ-REYES, R. (1971). Power and immortality: Essays on strategy, war psychology and war control. Jericho, N.Y.: Exposition Press.

LOVELL, J.P. (1964). "The professional socialization of the West Point cadet." Pp. 119-157 in M. Janowitz (ed.), The new military: Changing patterns of organizations. New York: Russell Sage Foundation.

LYON, W.B. (1967). Some reactions of children from military families to family mobility and father absence. Paper presented at the meeting of the American Psychological Association, Washington, D.C., September.

LYON, W.B., and OLDAKER, L. (1967). "The child, the school, and the military family." American Journal of Orthopsychiatry, 37:269-270.

MACINTOSH, H. (1968). "Separation problems of military wives." American Journal of Psychiatry, 125:260-265.

MAHAN, J.L., Jr., and CLUM, G.A. (1971). "Longitudinal prediction of Marine combat effectiveness." Journal of Social Psychology, 83:45-54.

MASLOW, A. (1954). Motivation and personality. New York: Harper and Row.

MASON, W.M. (1970). "On the socio-economic effects of military service." Unpublished Ph.D. dissertation, University of Chicago.

McCUBBIN, H., and DAHL, B. (1974a). "Factors in family reunion." Paper presented at the Prisoner of War Research Conference, Naval Health Research Center, San Diego, Calif., April.

——— (1974b). "Research on families of returned prisoners of war: A review." Paper presented at the second Joint Medical Meeting on Returned Prisoners of War, Department of Defense, San Antonio, Texas, November.

——— (1974c). "Social and mental health services to families of returned prisoners of war." Paper presented at the meeting of the American Psychiatric Association, Detroit, May.

McCUBBIN, H., DAHL, B., LESTER, G., and ROSS, B. (1975). "The returned prisoner of war: Factors in family adjustment." Paper presented at the International Conference on Psychological Stress and Adjustment in Time of War and Peace, Tel Aviv, Israel, January.

McCUBBIN, H., DAHL, B., METRES, P., HUNTER, E., and PLAG, J. (eds., 1975). Family separation and reunion: Families of prisoners of war and servicemen missing in action. Washington, D.C.: U.S. Government Printing Office.

McCUBBIN, H., HUNTER, E., and DAHL, B. (1975). "Residuals of war: Families of prisoners of war and servicemen missing in action." Journal of Social Issues, 31(4):91-110.

McDONALD, B.W., PUGH, W.M., and GUNDERSON, E.K.E. (1973). "Organizational factors and health status." Journal of Health and Social Behavior, 14:330-334.

McGOFF, R.M., and HARDING, F.D. (1974). "A report on literacy training programs in the armed forces." Alexandria, Va.: Air Force Human Resources Laboratory, Manpower Development Division.

McKAIN, J.L. (1969). "Feelings of alienation, geographical mobility and Army family problems: An extension of theory." Unpublished doctoral dissertation, Catholic University of America.

——— (1973). "Relocation in the military: Alienation and family problems." Journal of Marriage and Family, 35:205-209.

McNEIL, J.S. (1964). "Adjustment of retired Air Force officers." Unpublished doctoral dissertation, University of Southern California.

McNEIL, J.S., and GIFFEN, M. (1965a). "Military retirement: Some basic observations and concepts." Aerospace Medicine, 36:25-29.

--- (1965b). "The social impact of military retirement." Social Casework, 46:203-207.

--- (1967). "Military retirement: The retirement syndrome." American Journal of Psychiatry, 123:848-854.

McNEIL, J.S., and ZONDERVAN, R.L. (1971). "The family in cultural isolation." Military Medicine, 136:451-454.

MICHAELSEN, L.K. (1973). A methodology for the studies of the impact of organizational values, preferences, and practices on the all-volunteer Navy. Ann Arbor: Institute for Social Research, University of Michigan.

--- (1974). Attitudes and motivations toward enlistment in the U.S. Army. Princeton, N.J.: Opinion Research Corporation.

MILGRAM, R. (1975). "The effects of the Yom Kippur War on anxiety level in Israeli children." Paper presented at the International Conference on Psychological Stress and Adjustment in Time of War and Peace, Tel Aviv, Israel, January.

MILGRAM, S. (1963). "Behavioral study of obedience." Journal of Abnormal and Social Psychology, 67:371-378.

--- (1967). "The compulsion to do evil." Patterns of prejudice, 1(November-December).

--- (1974). Obedience to authority. New York: Harper and Row.

MILLER, A.H. (1974). "Political issues and trust in government: 1964-1970." American Political Science Review, 68(September):951-972.

MILLER, J.C., and TOLLISON, R. (1971). "The implicit tax on relevant military recruits." Social Science Quarterly, 51(March):924-931.

MILLER, T.W. (1973). "Peer counseling: A model for selection and training of students to help students." Counseling and Values, 17:190-194.

MILLS, C.W. (1956). The power elite. New York: Oxford University Press.

MILOWE, I. (1964). "A study in role diffusion: The chief and sergeant face retirement." Mental Hygiene, 48:101-107.

MODIGLIANI, A. (1972). "Hawks and doves, isolationism and political distrust: An analysis of public opinion on military policy. American Political Science Review, 66(September):960-978.

MONTALVO, F.F. (1968). "Family separation in the Army: A study of the problems encountered and the caretaking resources used by career Army families undergoing military separation." Unpublished doctoral dissertation, University of Southern California.

MOOS, R.H. (1974). "Psychological techniques in the assessment of adaptive behavior." Pp. 334-399 in G.V. Coelho, D.A. Hamburg, and J.E. Adams (eds.), Coping and adaptation. New York: Basic Books.

MOSKOS, C. C., Jr. (1970). The American enlisted man. New York: Russell Sage Foundation.

--- (1973a). "The emergent military: Civil, traditional, or plural?" Pacific Sociological Review, 16(April):255-280.

--- (1973b). "Studies on the American soldier: Continuities and discontinuities in social research." Paper presented at the meeting of the American Sociological Association, New York.

MURPHY, M.K., and PARKER, C.B. (1966). Fitting in as a new service wife. Harrisburgh, Pa.: Stackpole.

MYLES, D. (1970). "A survey of community services centers: Welfare problems, services, personnel and resources." Unpublished doctoral dissertation, Catholic University of America.

NELSON, P.D. (1971). "Personnel performance prediction." Pp. 91-121 in R.W. Little (ed.), Handbook of military institutions. Beverly Hills, Calif.: Sage.

——— (1974). "Comment: On life change and subsequent illness reports." Pp. 79-89 in E.K.E. Gunderson and R.H. Rahe (eds.), Life stress and illness. Springfield, Ill.: Charles C Thomas.

NELSON, P.D., and GUNDERSON, E.K.E. (1963). "Personal history correlates of performance among military personnel in Antarctic stations" (Report 63-20). San Diego, Calif.: U.S. Navy Medical Neuropsychiatric Research Unit.

NOWICKI, S., and STRICKLAND, B.R. (1973). "A locus of control scale for children." Journal of Consulting and Clinical Psychology, 40:148-154.

NUNNALLY, J.C. (1967). Psychometric theory. New York: McGraw-Hill.

OBRDLIK, A.J. (1943). "Gallows humor–A sociological phenomenon." American Journal of Sociology, 47(March):709-716.

O'GORMAN, J.G. (1972). "Review of research on re-engagement among Australian regular Army soldiers" (Research Report 2/72). Melbourne, Australia: 1 Psychological Research Unit.

Opinion Research Corporation (1974). Attitudes and motivation toward enlistment in the U.S. Army. Princeton, N.J.: Author.

OPPENHEIMER, M. (1971). "The military: What it does." Pp. 88-103 in M. Oppenheimer (ed.), The American Military. Chicago: Aldine.

OSBORN, W.C. (1973). "Developing performance tests for training evaluation" (HumRRO Professional Paper 3-73). Alexandria, Va.: Human Resources Research Office.

OWENS, A.G. (1968). "A further study of factors affecting the decision on reengagement among ARA soldiers." Australian Military Forces Research Report, pp. 1-68.

——— (1969). "Job satisfaction and reengagement among Australian regular army soldiers." Australian Journal of Psychology, 21:137-144.

OWENS, W.A. (1971). "A quasi-actuarial basis for individual assessment." American Psychologist, (November):992-999.

PALGI, P. (1973). Socio-cultural expressions and implications of death, mourning, and bereavement in Israel arising out of the war situation. Jerusalem: Jerusalem Academic Press.

——— (1975). "Culture and mourning: Expressions of bereavement arising out of the war situation in Israel." Paper presented at the International Conference on Psychological Stress and Adjustment in Time of War and Peace, Tel Aviv, Israel, January.

PATTERSON, R. (1945). "Neurotic reactions in wives of servicemen." Diseases of the Nervous System, 6:50-52.

PAYNE, G.D. (1974). "Measurement of the psychological characteristics of the environment of a military institution: A preliminary study" (Research Report 3/74). Melbourne, Australia: 1 Psychological Research Unit.

PEARLMAN, C.A. (1970). "Separation reactions of married women. American Journal of Psychiatry, 126:946-950.

PEDERSEN, F. (1966). "Relationship between father absence and emotional disturbance in male military dependents." Merrill-Palmer Quarterly, 12:321-333.

PEDERSEN, F., and SULLIVAN, E.J. (1964). "Relationships among geographical mobility, parental attitudes and emotional disturbances in children." American Journal of Orthopsychiatry, 34:575-580.

PEPIN, R. (1967). "A study of scholastic achievements and personal adjustments of military and non-military transitory high school students." Dissertation Abstracts, 28:403-404.

PICOU, J.S., and NYBERG, K.L. (1975). "Socialization into the military: A study of university cadets." Paper presented at the meeting of the American Sociological Association, San Francisco, August.

PILISUK, M., and HAYDEN, T. (1955). "Is there a military-industrial complex which prevents peace?" Journal of Social Issues, 21(July):67-117.

PLAG, J. (1974). "Proposal for the long-term follow-up of returned prisoners of war, their families, and the families of servicemen missing in action: A basis for the delivery of health care services." Paper presented at the POW Research Consultants Conference, San Diego, California, April.

PLAG, J., and GOFFMAN, J.M. (1966). "The prediction of four-year military effectiveness from characteristics of naval recruits." Military Medicine, 131:729-735.

--- (1969). "The utilization of predicted military effectiveness scores for selecting naval enlistees" (Report 69-6). San Diego, Calif.: U.S. Naval Medical Neuropsychiatric Research Unit.

--- (1973). "Characteristics of naval recruits with histories of drug abuse." Military Medicine, 138:354-359.

PLAG, J., and HARDACRE, L.E. (1964). "The validity of age, education, and GCT score as predictors of two-year attrition among naval recruits" (Report 64-15). San Diego, Calif.: U.S. Navy Medical Neuropsychiatric Research Unit.

PLATTE, R. (1974). "The second career: Perceived social mobility and adjustment among recent Army retirees and wives of Army retirees." Unpublished doctoral dissertation, Florida State University.

POLNER, M. (1971). No victory parades: The return of the Vietnam veteran. New York: Holt, Rinehart and Winston.

PRICE-BONHAM, S. (1970). "A study of thirty-two wives whose husbands are missing in action in Vietnam." Paper presented at the National Council on Family Relations, Chicago, October.

PUGH, W.M., and GUNDERSON, E.K.E. (1975). "Individual and situational predictors of illness" (Report 75-20). San Diego, Calif.: Naval Health Research Center.

RADLOFF, R., and HELMREICH, R. (1968). Groups under stress. New York: Appleton-Century-Crofts.

RAHE, R.H. (1972). "Subjects' recent life changes and their near-future illness reports." Annals of Clinical Research, 4:250-265.

RAHE, R.H., GUNDERSON, E.K.E., and ARTHUR, R.J. (1970). "Demographic and psychosocial factors in acute illness reporting." Journal of Chronic Disease, 23:245-255.

RAHE, R.H., GUNDERSON, E.K.E., PUGH, W.M., RUBIN, R.T., and ARTHUR, R.J. (1972). "Illness prediction studies: Use of psychological and occupational characteristics as predictors." Archives of Environmental Health, 25:192-197.

RAHE, R.H., MAHAN, J., ARTHUR, R.J., and GUNDERSON, E.K.E. (1970). "Epidemiology of illness in naval environments. I. Illness types, distribution, severities, and relationship to life change." Military Medicine, 135: 443-452.

RAKOFF, V. (1966). "A long-term effect of the concentration camp experience." Viewpoints, 1:17-22.

RAKOFF, V., SIGAL, J., and EPSTEIN, N. (1967). "Children and families of concentration camp survivors." Canadian Mental Health, 14:24-26.

RIMLAND, B., and NEUMANN, I. (1966). "A two-year follow-up of Marines assigned to a special (correctional) training branch during recruit training" (Report 67-3). San Diego, Calif.: U.S. Naval Personnel Research Activity.

RIOCH, D.M. (1968). "Prevention: The major task of military psychiatry." Psychotherapy and Psychosomatics, 16:55-63.

RODGERS, W.L., and JOHNSTON, L.D. (1974). "Attitudes toward business and other American institutions." Paper presented at the Conference of the American Association for Public Opinion Research, San Francisco, May.

ROE, F.M.A. (1909). Army letters from an officer's wife, 1871-1888. New York: D. Appleton.

ROFF, M.E. (1950). "A study of combat leadership in the Air Force by means of a rating scale." Journal of Psychology, 30:229-239.

ROGERS, C. (1944). "Wartime issues in family counseling." Marriage and Family Living, 68-69:84.

ROGHMANN, K. (1966). Dogmatismus and Authoritarismus. Meissenheim, W. Ger.: Anton Hain.

ROGHMANN, K., and SODEUR, W. (1972). "The impact of military service on authoritarian attitudes." American Journal of Sociology, 78(September):418-433.

--- (1973). "Reply to Stinchcombe." American Journal of Sociology, 79:159-164.

ROKEACH, M. (1968). Beliefs, attitudes and values. San Francisco: Jossey-Bass.

--- (1973). The nature of human values. New York: Free Press.

ROSSI, A.S. (1972). "Family development in a changing world." American Journal of Psychiatry, 128:1057-1066.

RUBIN, R.T., GUNDERSON, E.K.E., and ARTHUR, R.J. (1969). "Life stress and illness patterns in the U.S. Navy: III. Prior life change and illness onset in an attack carrier's crew." Archives of Environmental Health, 19:753-757.

--- (1971a). "Life stress and illness patterns in the U.S. Navy: IV. Environmental and demographic variables in relation to illness onset in a battleship's crew." Journal of Psychosomatic Research, 15:277-288.

--- (1971b). "Life stress and illness patterns in the U.S. Navy: V. Prior life change and illness onset in a battleship's crew." Journal of Psychosomatic Research, 15:89-94.

RUBIN, R.T., GUNDERSON, E.K.E., and DOLL, R.E. (1969). "Life stress and illness patterns in the U.S. Navy: I. Environmental variables and illness onset in an attack carrier's crew." Archives of Environmental Health, 19:740-747.

RUE, V.M. (1973). "A U.S. Department of Marriage and the family." Journal of Marriage and the Family, 35:689-699.

RYAN, F., and BEVILACQUA, J. (1964). "The military family: An asset or a liability." Military Medicine, 129:956-959.

RYDER, N.B. (1974). "The family in developed countries." Scientific American, 231:122-132.

SAKLEM, A.A., CASTLE, J.E., and WEILER, D.J. (1971). "The shipboard environment –Past, present, and future." Naval Engineers Journal, 83:102-113.

SALAS, R.G. (1968). "The assimilation of male volunteer recruits to the Australian regular army: A study of the process of militarisation (4 parts)" (Research Reports 8-11/68). Melbourne, Australia: 1 Psychological Research Unit.

--- (1974). "The expectations, social attitudes, values, motivations, interests and satisfactions of Australian army officer cadets" (Research Report 5/73). Melbourne, Australia: 1 Psychological Research Unit.

SANUA, V. (1974). "The psychological effects of the Yom Kippur War." Unpublished manuscript, City College, New York.

SARKESIAN, S. (ed., 1972). The military-industrial complex: A reassessment. Beverly Hills, Calif.: Sage.

SARKESIAN, S., and TAYLOR, W.J., Jr. (1975). "The case for civilian graduate education for professional officers." Armed Forces and Society, 1(winter):251-262.

SATTIN, D., and MILLER, J. (1971). "The ecology of child abuse within a military community." American Journal of Orthopsychiatry, 41:675-678.

SAUNDERS, D. (1969). "Poverty in Army families." Social Service Review, 43:96-99.

SCHNEIDER, D.M. (1953). "The social dynamics of physical disability in Army basic training." Pp. 386-397 in C. Kluckholm and H.A. Murray (eds.), Personality in nature, society and culture. New York: Knopf.

SCHUETZ, A. (1945). "The homecomer." American Journal of Sociology, 10:369-376.

SCHUMAN, H. (1972). "Two sources of antiwar sentiment in America." American Journal of Sociology, 78(November):513-536.

SEGAL, D.R. (1975a). "Civil-military relations in the mass public." Armed Forces and Society, 1:215-229.

--- (1975b). "Convergence, commitment and military compensation." Paper delivered at the meeting of the American Sociological Association, San Francisco, August.

--- (1975c). "Public opinion on the military establishment and military policy." Paper prepared for the 1975 meeting of the British Inter-University Seminar on Armed Forces and Society, Cambridge University, April.

SEGAL, D.R., BLAIR, J., NEWPORT, F., and STEPHENS, S. (1974). "Convergence, isomorphism and interdependence at the civil-military interface." Journal of Political and Military Sociology, 2(fall):157-172.

SEGAL, J. (1973). "Therapeutic considerations in planning the return of American POWs to continental United States." Military Medicine, 138:73-77.

SEIDENBERG, R. (1973). Corporate wives-corporate casualties. New York: AMACOM.

SELLS, S.B., and GUNDERSON, E.K.E. (1972). "A social system approach to long-duration missions." In Space Science Board (ed.), Human factors in long-duration spaceflight. Washington, D.C.: National Academy of Science.

SEPLIN, C. (1952). "A study of the influence of father's absence for military service." Smith College Studies in Social Work, 22:123-124.

SHARP, L.M. (1970). "The role of military service in the careers of college graduates." Chap. 5 in L.M. Sharp (ed.), Education and employment: The early careers of college graduates. Baltimore: Johns Hopkins University Press.

SHATAN, C.F. (1973). "How do we turn off the guilt?" Human Behavior, 2:56-61.

SHAW, J.A., and PANGMAN, J. (1975). "Geographic mobility and the military child." Military Medicine, 140:413-416.

SHEA, N. (1954). The Army wife. New York: Harper and Row.

SHILS, E.A., and JANOWITZ, M. (1948). "Cohesion and disintegration in the Wehrmacht in World War II." Public Opinion Quarterly, 12(summer):280-315.

SIGAL, J. (1971). "Second generation effects of massive psychiatric trauma." International Psychiatric Clinic, 8:55-65.

SIGAL, J., and RAKOFF, B. (1971). "Concentration camp survival: A pilot study of effects on the second generation." Canadian Psychiatric Association Journal, 16:393-397.

SIGAL, J., SILVER, V., RAKOFF, B., and ELLIN, B. (1973). "Some second-generation effects of survival of the Nazi persecution." American Journal of Orthopsychiatry, 43:320-326.

SIRJAMAKI, J. (1964). "The institutional approach." Pp. 33-50 in H. Christensen (ed.), Handbook of marriage and the family. Chicago: Rand McNally.

SMILANSKY, S. (1975). "Development of the conceptualization of death in children, ages 4-10." Paper presented at the International Conference on Psychological Stress and Adjustment in Time of War and Peace, Tel Aviv, Israel, January.

SOLOMON, G.F., ZARCONE, V.P., YOERG, R., SCOTT, N.R., and MAURER, R.G. (1971). "Three psychiatric casualties from Vietnam." Archives of General Psychiatry, 25:522-524.

SPELLMAN, S. (1965). "Orientations toward problem-solving among career military families." Unpublished doctoral dissertation, School of Social Work, Columbia University.

STANTON, M.D. (1973). "The soldier." Chap. 21 in D. Spiegel and P.K. Spiegel (eds.), Outsiders, USA. San Francisco: Rinehart.

--- (1975). "Psychology and family therapy." Professional Psychology, 6:45-49.

STARR, P. (1973). The discharged Army: Veterans after Vietnam. New York: Charterhouse.

STEIGER, W.A. (1971). "Welfare, extra jobs sustain GI families." Washington Post, May 9.

STERNBERG, T. (1975). "Work of volunteers with bereaved families of the Yom Kippur War." Paper presented at the International Conference on Psychological Stress and Adjustment in Time of War and Peace, Tel Aviv, Israel, January.

STINCHCOMBE, A. (1973). "Comment on 'the impact of military service on authoritarian attitudes'." American Journal of Sociology, 79:157-159.

STODDARD, E.R. (1954). "The military intelligence agent: Structural strains in an occupational role." Pp. 570-582 in C.D. Bryant (ed.), Social dimensions of work. Englewood Cliffs, N.J.: Prentice-Hall.

STOKOLS, D. (1972). "On the distinction between density and crowding: Some implications for future research." Psychological Review, 79:275-277.

STOLZ, L.M. (1954). Father relations of war-born children. Palo Alto, Calif.: Stanford University Press.

STOUFFER, S.A., DeVINNEY, L.C., STAR, S.A., and WILLIAMS, R.M. (1949). The American soldier: Adjustment during Army life. Princeton, N.J.: Princeton University Press.

STOUFFER, S.A., LUMSDAINE, A.A., LUMSDAINE, M.H., WILLIAMS, R.M., SMITH, M.B., JANIS, I.L., STAR, S.A., and COTRELL, L.S. (1949). The American soldier: Combat and its aftermath. Princeton, N.J.: Princeton University Press.

STRANGE, R.E., and ARTHUR, R.J. (1967). "Hospital ship psychiatry in a war zone." American Journal of Psychiatry, 124:281-286.

STRANGE, R.E., and BROWN, D.E. (1970). "Home from the war: A study of psychiatric problems in Vietnam returnees." American Journal of Psychiatry, 127:488-492.

SUBRAMIAM, P., and OWENS, A.G. (1974). "The satisfactions and dissatisfactions of medical officers in the services" (Research Report 2/74). Melbourne, Australia: 1 Psychological Research Unit.

SULLIVAN, J.A. (1970). "Qualitative requirements of the armed forces." Studies prepared for the President's Commission on an All-Volunteer Armed Force I (November).

SUMMERHAYES, M. (1911). Vanished Arizona. Salem, Mass.: Salem Press.

TAX, S. (ed., 1967). The draft: A handbook of facts and alternatives. Chicago: University of Chicago Press.

TAYLOR, J.C., and BOWERS, D.G. (1972). Survey of organizations: A machine-scored standardized questionnaire instrument. Ann Arbor: Institute for Social Research, University of Michigan.

TEICHMAN, Y., and SPIEGEL, Y. (1975). "Volunteers report about their work with families of servicemen missing in action." Paper presented at the International Conference on Psychological Stress and Adjustment in Time of War and Peace, Tel Aviv, Israel, January.

TOBIN, D.J. (1967). "Performance appraisal by the cadet chain of command." West Point, N.Y.: Office of Military Psychology and Leadership, U.S. Military Academy.

TOBIN, D.J., and MARCRUM, R.H. (1967). "Leadership evaluation." West Point, N.Y.: Office of Military Psychology and Leadership, U.S. Military Academy.

TROSSMAN, B. (1968). "Adolescent children of concentration camp survivors." Canadian Psychiatric Association Journal, 13:121-123.

TRUNNELL, T. (1968a). "The absent father's children's emotional disturbances." Archives of General Psychiatry, 19:180-183.

--- (1968b). "A review of studies of the psychosocial significance of the absent father." Paper presented at the meeting of the American Psychiatric Association, Boston, May.

TURNER, R.H. (1947). "The Navy disbursing officer as a bureaucrat." American Sociological Review, 12(June):342-348.

TUTTLE, T.C., and HAZEL, J.T. (1974). "Review and implications of job satisfaction and work motivation theories for Air Force research" (U.S. AFHRL-TR-73-56). San Antonio, Texas: Lackland Air Force Base.

TYHURST, J.S. (1957). "The role of transition states–including disasters–in mental illness." Pp. 149-169 in Symposium on Preventive and Social Psychiatry, Walter Reed Army Institute of Research. Washington, D.C.: U.S. Government Printing Office.

U.S. Army Times Publishing Co. (eds., 1974). Military Market Facts Book, 1974. Washington, D.C.: Author.

U.S. Bureau of the Census (1974). Statistical Abstract of the United States, 1974. Washington, D.C.: U.S. Government Printing Office.

U.S. Congress (1889-1890). Report of the Surgeon-General. 42nd Congress, 2nd Session (House Executive Document 1, pt. 2, II).

––– (1890-1891). Report of the Surgeon General. 51st Congress, 2nd Session (House Executive Document 1, pt. 2, II).

––– (1892). Senate Report 268. 52nd Congress, 1st Session.

––– (1908). "Educating children at military posts." Senate Document 349.

U.S. Department of the Army (1971). Leadership for the 1970's. Carlisle Barracks, Pa.: U.S. Army War College.

U.S. Department of Defense (1974). "Military Manpower Training Report for FY 1975" (March).

U.S. President's Commission on an All-Volunteer Armed Force (1970). The report of the President's Commission on an All-Volunteer Armed Force. Washington, D.C.: U.S. Government Printing Office.

U.S. Senate (1970). "Oversight of medical care of veterans wounded in Vietnam." In Hearings Before the Subcommittee on Veterans' Affairs of the Committee on Labor and Public Welfare, United States Senate. Washington, D.C.: U.S. Government Printing Office.

––– (1971). "Unemployment and overall readjustment problems of returning veterans." In Hearings Before the Subcommittee on Veterans' Affairs of the Committee on Labor and Public Welfare, United States Senate. Washington, D.C.: U.S. Government Printing Office.

U.S. War Department (1947). General regulations for the Army, 1847. Washington, D.C.: U.S. Government Printing Office.

U.S. War Office (1917). Regulations for the Army of the United States, 1913 corrected to 1917. Washington, D.C.: U.S. Government Printing Office.

USEEM, M. (1973). Conscription, protest, and social conflict. New York: John Wiley.

VERBA, S., and NIE, N. (1973). Participation in America. New York: Harper and Row.

VIDICH, A., and STEIN, M. (1960). "The dissolved identity in military life." Pp. 493-505 in M. Stein, A. Vidich, and D. White (eds.), Identity and anxiety. Glencoe, Ill.: Free Press.

VINCENT, C.E. (1970). "Mental health and the family." In P.H. Glasser and L.N. Glasser (eds.), Families in crisis. New York: Harper and Row.

VINEBERG, R., and TAYLOR, E.N. (1972). Summary and review of studies of the VOLAR experiment, 1971. (HumRRO) Technical Report 72-18). Alexandria, Va.: Human Resources Research Organization.

WALLER, W. (ed., 1940). War and the family. New York: Dryden.

––– (1944). The veteran comes back. New York: Dryden.

WEBSTER, J.C. (1975). "Noise and ship habitability." San Diego, Calif.: Naval Electronics Laboratory Center.

WEIDER, S., and NASHIM, E. (1975). "Parallel reactions of widows and young children following the Yom Kippur War." Paper presented at the International Conference on Psychological Stress and Adjustment in Time of War and Peace, Tel Aviv, Israel, January.

WEIGLEY, R.F. (1972). "A historian looks at the Army." Military Review, 51(February): 25-34.

WEISEL, V. (1967). Lady in arms. Boston: Houghton Mifflin.

WESTLING, L., Jr. (1973). "Ministry to prisoner of war returnees in the long-term readjustment period: A manual for Navy chaplains." Unpublished doctoral dissertation, U.S. Navy Post-Graduate School.

WHYTE, W.H. (1956). The organization man. New York: Simon and Schuster.

WILKINS, W.L. (1961). "The identification of character and behavior disorders in the military life." Pp. 663-680 in Fifth Navy Science Symposium (ONR-9). Washington, D.C.: U.S. Office of Naval Research.

WILLIAMS, J. (1974). "Divorce and the Air Force family." Air Force Magazine, October.

WILSON, S.R., and FLANAGAN, J.C. (1974). Supplementary report: Quality of life as perceived by 30 year old Army veterans. Palo Alto, Calif.: American Institutes for Research.

WILSON, S.R., FLANAGAN, J.C., and UHLANER, J.E. (1975). Quality of life as perceived by 30 year old Army veterans (Technical Paper). Arlington, Va.: U.S. Army Research Institute for the Behavioral and Social Sciences.

WOOD, B.D. et al. (1953). "Survey of the aptitude for the service rating system at the United States Military Academy." West Point, N.Y.: U.S. Military Academy.

WOOL, H. (1968). The military specialist: Skilled manpower for the armed forces. Baltimore: Johns Hopkins University Press.

WOOSTER, A.D., and HARRIS, G. (1972). "Concepts of self and others in highly mobile service boys." Educational Research, 14:195-199.

YARMOLINSKY, A. (1971). The military establishment. New York: Harper and Row.

--- (1972). The military establishment: Its impacts on American society. New York: Harper and Row.

ZIV, A. (1975). "Empirical findings on children's reactions to war stress." Paper presented at the International Conference on Psychological Stress and Adjustment in Time of War and Peace, Tel Aviv, Israel, January.

ZUNIN, L. (1969). "Second life for war widows." Time, 9:54-55.

ZUNIN, L., and BARR, N. (1969). "Therapy program aids servicemen's widows." U.S. Medicine, 5:12.

NOTES ON THE CONTRIBUTORS

JERALD G. BACHMAN is a Program Director and Research Scientist at the University of Michigan's Institute for Social Research. His research in the Youth in Transition project, and Monitoring the Future, and his collaboration with David G. Bowers have contributed to his continued interest in military staffing under all-volunteer conditions.

JOHN D. BLAIR was Assistant Study Director at the Survey Research Center of the Institute for Social Research, University of Michigan. Currently he is Assistant Professor in the Department of Sociology at the University of Maryland. He and Jerald G. Bachman have recently completed a monograph entitled *Soldiers, Sailors and Civilians: The "Military Mind" and the All Volunteer Force.*

JONATHAN F. BORUS is an Assistant Professor of Psychiatry at the Harvard Medical School. He is the Director of Training at the Erich Lindemann Mental Health Center in Boston and Coordinator of Social and Community Psychiatry Training for the Massachusetts General Hospital. His current research includes a focus on the transition periods in the personal and professional lives of physicians.

DAVID G. BOWERS is Program Director of the Center for Research on Utilization of Scientific Knowledge, Institute for Social Research, University of Michigan. He has recently been involved in studies of the organizational climate of the Army and Navy.

CLAUDE E. CABANILLAS is a Captain in the Signal Corps, U.S. Army. He is an instructor of Leadership and Human Relations with the Department of Leadership and Resource Management, United States Army Sergeants Major Academy, Fort Bliss, Texas.

JOHN H. FARIS is Assistant Professor of Sociology at Towson State University. "The United States Army Basic Combat Training in the

Transition from Conscription to the All-Volunteer Force," the title of his doctoral dissertation, has been the focus of his research.

NANCY L. GOLDMAN is a Research Associate at the University of Chicago. In addition to her interest in the military family, she has been involved in research on the utilization of women in the military.

E.K. ERIC GUNDERSON is head of Environmental and Social Medicine at the Naval Health Research Center, San Diego, and Adjunct Professor of Psychiatry, University of California, San Diego. He directs a broad research program concerned with environmental and social factors in disease and behavior pathology.

RONALD INGLEHART is Associate Professor of Political Science, University of Michigan, where he is also a Research Associate in the Center for Political Studies, Institute for Social Research. Since 1970, during each spring term, he has been visiting professor in the Department of Political Science at the University of Geneva, Switzerland.

M. KENT JENNINGS is Professor of Political Science, a Program Director of the Center for Political Studies, and Associate Director of the Inter-University Consortium for Political Research, University of Michigan. His major areas of interest are political socialization, electoral behavior and public opinion, and community politics.

ROBERT JAY LIFTON is Professor in the Department of Psychiatry at the Yale University School of Medicine. His research interests have included the psychological aspects of totalitarianism and of nuclear weapons as well as the adjustment of the Vietnam veteran.

GREGORY B. MARKUS is an Assistant Research Scientists at the Institute for Social Research and an Assistant Professor of Political Science at the University of Michigan. He is presently engaged in research on generational change in political orientations.

PAUL D. NELSON, a Commander in the Medical Service Corps, U.S. Navy, is head of the Human Performance Research Division, Naval Medical Research and Development Command, Bethesda, Maryland.

As such he is responsible for management of research conducted through the Navy Medical Department on psychiatric adjustment to service, human behavior and health, and performance effectiveness under stresses of operational duties and naval environments.

DAVID R. SEGAL is Professor of Sociology and of Government and Politics at the University of Maryland. He is currently completing a monograph on civil-military relations and is engaged in research on human resource management in military organizations.

MADY WECHSLER SEGAL is Assistant Professor of Sociology at the University of Maryland. She has been involved in research in the role of women in the military and in studies of military families.

M. DUNCAN STANTON is Assistant Professor of Psychology in Psychiatry at the University of Pennsylvania and Director of the Addicts and Families Project at the Philadelphia Child Guidance Clinic. His recent research focuses on the family characteristics of drug-addicted veterans and the use of family therapy in their rehabilitation.

ELLWYN R. STODDARD is Professor of Sociology and Anthropology at the University of Texas, El Paso. The Army professionalization process and the social organization of the Army in an era of technological change are among his research concerns.